Dominique Manuel Schneuwly

Spatio-temporal rockfall analysis using tree ring data

Dominique Manuel Schneuwly

Spatio-temporal rockfall analysis using tree ring data

Anatomic tree reactions and spatio-temporal rockfall analysis - an applied study based on dendrogeomorphic methods

Südwestdeutscher Verlag für Hochschulschriften

Impressum/Imprint (nur für Deutschland/ only for Germany)
Bibliografische Information der Deutschen Nationalbibliothek: Die Deutsche Nationalbibliothek verzeichnet diese Publikation in der Deutschen Nationalbibliografie; detaillierte bibliografische Daten sind im Internet über http://dnb.d-nb.de abrufbar.

Alle in diesem Buch genannten Marken und Produktnamen unterliegen warenzeichen-, marken- oder patentrechtlichem Schutz bzw. sind Warenzeichen oder eingetragene Warenzeichen der jeweiligen Inhaber. Die Wiedergabe von Marken, Produktnamen, Gebrauchsnamen, Handelsnamen, Warenbezeichnungen u.s.w. in diesem Werk berechtigt auch ohne besondere Kennzeichnung nicht zu der Annahme, dass solche Namen im Sinne der Warenzeichen- und Markenschutzgesetzgebung als frei zu betrachten wären und daher von jedermann benutzt werden dürften.

Verlag: Südwestdeutscher Verlag für Hochschulschriften Aktiengesellschaft & Co. KG
Dudweiler Landstr. 99, 66123 Saarbrücken, Deutschland
Telefon +49 681 37 20 271-1, Telefax +49 681 37 20 271-0
Email: info@svh-verlag.de
Zugl.: Fribourg, University of Fribourg, Dissertation, 2009

Herstellung in Deutschland:
Schaltungsdienst Lange o.H.G., Berlin
Books on Demand GmbH, Norderstedt
Reha GmbH, Saarbrücken
Amazon Distribution GmbH, Leipzig
ISBN: 978-3-8381-1291-6

Imprint (only for USA, GB)
Bibliographic information published by the Deutsche Nationalbibliothek: The Deutsche Nationalbibliothek lists this publication in the Deutsche Nationalbibliografie; detailed bibliographic data are available in the Internet at http://dnb.d-nb.de.

Any brand names and product names mentioned in this book are subject to trademark, brand or patent protection and are trademarks or registered trademarks of their respective holders. The use of brand names, product names, common names, trade names, product descriptions etc. even without a particular marking in this works is in no way to be construed to mean that such names may be regarded as unrestricted in respect of trademark and brand protection legislation and could thus be used by anyone.

Publisher: Südwestdeutscher Verlag für Hochschulschriften Aktiengesellschaft & Co. KG
Dudweiler Landstr. 99, 66123 Saarbrücken, Germany
Phone +49 681 37 20 271-1, Fax +49 681 37 20 271-0
Email: info@svh-verlag.de

Printed in the U.S.A.
Printed in the U.K. by (see last page)
ISBN: 978-3-8381-1291-6

Copyright © 2010 by the author and Südwestdeutscher Verlag für Hochschulschriften Aktiengesellschaft & Co. KG and licensors
All rights reserved. Saarbrücken 2010

TABLE OF CONTENTS

TABLE OF CONTENTS ———————————— 1
ABSTRACT ———————————————— 3
ZUSAMMENFASSUNG ————————————— 7
ACKNOWLEDGEMENTS ————————————— 11

CHAPTER A: GENERAL INTRODUCTION

1 INTRODUCTION ———————————— 15

1.1 BACKGROUND ———————————— 15
1.2 AIMS OF THE STUDY —————————— 16
1.3 STRUCTURE OF THE THESIS ——————— 17

2 UNDERSTANDING ROCKFALL ———————— 19

2.1 INTRODUCTION ———————————— 19
   2.1.1 DEFINITIONS —————————— 19
   2.1.2 CLASSIFICATION ————————— 19
2.2 FROM TOP TO BOTTOM ————————— 21
   2.2.1 PROMOTERS AND TRIGGERS —————— 21
   2.2.2 ZONES OF A ROCKFALL SLOPE ————— 23
   2.2.3 ROCKS MOVING —————————— 24
   2.2.4 ROCKFALL VELOCITIES AND REBOUND
        PARAMETERS —————————— 26
2.3 HAZARD ANALYSIS, RISK ASSESSMENT —— 27
2.4 MODELING ————————————— 27
   2.4.1 EMPIRICAL ROCKFALL MODELS ———— 28
   2.4.2 PROCESS-BASED ROCKFALL MODELS — 28
   2.4.3 GIS-BASED ROCKFALL MODELS ———— 28
2.5 ROCKFALL RECONSTRUCTIONS ————— 29

3 ROCKFALL-FOREST INTERACTIONS ———— 31

3.1 INFLUENCE OF TREES ON ROCKFALL —— 31
   3.1.1 DEPARTURE ZONE ————————— 31
   3.1.2 TRANSITION AND DEPOSITION ZONE — 32
3.2 INFLUENCE OF ROCKFALL ON TREES —— 34

4 DENDROGEOMORPHOLOGY ——————— 37

4.1 BACKGROUND ———————————— 37
4.2 TREES AND ROCKFALL ————————— 38
   4.2.1 WOUNDING (INJURIES) ——————— 38
   4.2.2 BREAK-OFF OF BRANCHES AND STEM
        BREAKAGE ——————————— 39
   4.2.3 INCLINATION OF STEM ——————— 40
   4.2.4 ELIMINATION OF NEIGHBORING TREES — 41
4.3 METHODS IN DENDROGEOMORPHOLOGY — 42
   4.3.1 LOCALIZATION —————————— 42
   4.3.2 SAMPLING METHODS AND STRATEGIES — 42
   4.3.3 SAMPLING PREPARATION AND ANALYSIS — 46
   4.3.4 DATING ROCKFALL EVENTS —————— 47
4.4 DENDROGEOMORPHIC RESEARCH – A
     SHORT OVERVIEW —————————— 47

5 STUDY SITES —————————————— 49

5.1 SAAS BALEN ————————————— 50
5.2 TÄSCH ——————————————— 51
5.3 VAUJANY —————————————— 52

6 BIBLIOGRAPHY OF CHAPTER A ————— 53

Chapter B: Fundamental Research

1 Formation and spread of callus tissue and tangential rows of resin ducts in *Larix decidua* and *Picea abies* following rockfall impacts 69

2 Three-dimensional analysis of the anatomical growth response of European conifers to mechanical disturbance 83

Chapter C: Applied Research

1 Tree-ring based reconstruction of the seasonal timing, major events and origin of rockfall on a case-study slope in the Swiss Alps 103

2 Spatial analysis of rockfall activity, bounce heights and geomorphic changes over the last 50 years – A case study using dendrogeomorphology 119

Chapter D: Synthesis

1 Overall discussion and conclusions 141

  1.1 Main results 141
  1.2 Dendrogeomorphology in rockfall research 143
    1.2.1 *Potential* *143*
    1.2.2 *Limitations* *143*
  1.3 Implications of results 144
    1.3.1 *Rockfall triggers* *144*
    1.3.2 *Rockfall and climate change* *146*
    1.3.3 *Future rockfall at the study site* *147*
    1.3.4 *Forest protection effect* *148*
  1.4 Final considerations 148

2 Remaining questions and further research 149

3 Bibliography of Chapter D 151

♦♦♦♦♦

Abstract

Geomorphic processes continuously shape mountain regions with rockfall being one of the most common and frequent processes. In contrast to other processes, such as glaciations, flooding, or debris flow events, regular rockfall events do not have the similar potential in shaping entire mountain regions. Rockfall in general stands out more because of its high frequency than its devastating magnitude.

However, the unpredictable and sudden occurrence of rockfall poses a major threat to settlements, human infrastructure and can even lead to loss of life. The expansion of anthropogenic activities into marginal regions as well as the trend towards more outdoor-activities increase the probability of rockfall accidents. Understanding of the process and accurate hazard assessment is inevitable in order to avoid future damage.

As a result of the sudden occurrence and unpredictability of rockfall events, they can rarely be observed or recorded. Therefore, determining specific rockfall properties poses major difficulties. As real-time observations of rockfall are not suitable for data gathering, the reconstruction of historic events remains as an alternative data source. However, the perusal of archival data on past events remains scarce and, if available, fragmentary. Events were only noted if they caused severe damage to infrastructure or led to loss of life. Data on small-scale events and even more on events in uninhabited areas are generally missing. All methods used so far for rockfall reconstruction exclusively investigated rockfall deposits and yielded at most general and unspecific data. Sophisticated rockfall models attempt to simulate rockfall events and delimit endangered areas. However, obtained model predictions need to be verified with real data. Therewith, an inaccurate model can be calibrated and differences between model results and reality minimized. This procedure requires accurate data on past rockfall events.

The remaining alternative in order to reconstruct rockfall events is dendrogeomorphology. It is the only known method that has the potential to localize and date past events with yearly precision as well as to obtain *in situ* data on numerous rockfall parameters, such as source area, trajectories, frequency, magnitude, seasonality, or identification of triggers. Therefore, it was the aim of this study to refine dendrogeomorphic methods and to apply them to several case-study sites. The goal of the two first studies was to assess the anatomical reactions of a tree following rockfall impact. Gathered knowledge was used in the two later studies to show possible applications.

4 - Abstract

In the first study, cross-sections from 111 tress severely damaged by rockfall (*Larix decidua* Mill. [European larch] and *Picea abies* (L.) Karst [Norway spruce]) were investigated so as to quantify the tangential extend of anatomic tree response at the height of wounding. Results reveal that trees react with the formation of callus tissue and especially with the formation of tangential rows of traumatic resin ducts (TRD). *Larix decidua* trees form callus tissue on 4.1% of the cross-section where the cambium has not been destroyed, TRD can be found on 34%. *Picea abies* shows the same anatomic responses and similar extent with callus tissue on 3.5% and TRD on 36.2%. Long-term production of anatomic growth features however reveals distinct differences between the species. Three years after wounding, TRDs can be found in 10% of *Larix decidua* trees. More than 50% of *Picea abies* individuals, however, produce TRDs even five years after impact. In addition, a shift in seasonal appearance of TRD was detected. With increasing distance from the wound, a delayed formation of TRD could be observed. In a last analytical step, the intensity of reactions was opposed to several tree parameters. Results indicate a weak positive correlation between the tangential extent and width of the wound. A slightly negative correlation was found between injury width and tree age or tree diameter. Younger individuals with a smaller diameter form less TRD than older individuals with a larger diameter.

The second paper investigated anatomic tree reactions following rockfall impact as well. Analysis however was not limited to the tangential extent, as tree reactions were additionally assessed in the axial direction. The focus was on the arrangement and occurrence probability of TRD and on the formation of reaction wood. In total, 16 entire trees (*Larix decidua*, *Picea abies* and *Abies alba* Mill. [Silver fir]) with 54 injuries were cut into segments of 10 cm height, resulting in 820 cross-sections. Results indicate that *Picea abies* trees form tangentially and axially more extended TRD than *Larix decidua*. In *A. alba*, in contrast, TRD can be found only sparsely. TRD can be found at highest probability in a position close above the wound. At this spot, TRD in *Larix decidua* and *Picea abies* can be found in more than 90% of the cases. In general, more TRD are present above than below the wound.

No correlation exists between the width or size of the wound and the spread of TRD. Analysis of long-term appearance of TRD following wounding shows that TRD are formed for most consecutive years in *Abies alba* (3.5 years), followed by *Larix decidua* and *Picea abies* (each 1.5 years). Tree-ring analyses furthermore reveal that additionally to the tangential seasonal delay of TRD formation, a similar shift exists in the vertical direction. A vertical shift can be found in all species, the shifting rate does not differ significantly. Another tree response to mechanical disturbance is differentiation of reaction wood. If a tree was tilted by an impact, it may regain its vertical stability through the formation of reaction wood. The presence of reaction wood strongly depends on the tree species. In *Larix decidua* trees, reaction wood could be found in only 8% of all injuries. In contrast, it was present in 50% of the *Picea abies* and even on 88% of the *Abies alba* injuries.

Based on the results obtained in the two theoretical papers, several applications were demonstrated in the third article. At a rockfall slope near Saas Balen (Valais, Switzerland), 154 wounds of 32 trees damaged by rockfall were investigated (*Larix decidua*, *Picea abies*, *Pinus cembra* L. [Swiss stone pine]). Existing growth features were used to precisely date each injury. The resulting reconstruction of five decades of past rockfall frequency indicates no increase in activity due to effects of the ongoing climate change. It was shown that 76% of all injuries occurred outside the vegetation period (early October – end May), whereas only 2.6% happened between mid July and early October. Data indicate that a combination of freeze-thaw related processes and precipitation (snow and rain) in spring time trigger most rockfall events. Additionally, two years with major rockfall activity could be assessed (1995, 2004). In both years, exceptionally heavy rainfalls were recorded in late autumn and spring respectively. The investigation on the orientation of the wounds finally allowed for a determination of the main rockfall source area.

More practical applications were presented in the fourth article. In total, 167 trees (*Larix decidua*, *Picea abies*, and *Pinus cembra*) were sampled on a rockfall slope near Saas Balen (Valais, Switzerland). The resulting 937 samples yielded data on 650 rockfall events. Again, no increase in rockfall activity could be identified for the last 50 years. In a first step, the afforestation process was reconstructed. Trees firstly appeared in the early 1950s at the lowest central part of the study site before they gradually moved up during the following years. The analysis of the injury heights revealed that the bounce heights of falling boulders averages 85 cm. Approximately two thirds of all injuries occur below 1 m, maximum recorded bounce height was 4.5 m. The large dataset allowed for a spatial analysis of bounce height, rockfall activity, and general growing conditions. Highest rebound heights occur at the lateral boundaries of the forest, lowest values in the central part. This finding can be explained by the shielding effect of an enormous boulder in the central top position of the study area and by the protection effect of the forest itself further downslope. Rockfall activity is highest at the lateral boundary of the forest that is oriented towards the main rockfall source area. Inside the forest, rockfall rates rapidly decrease, confirming the protective effect of the forest. Analysis of aerial photographs revealed the detachment of the immense boulder above the study area sometimes between 1958 and 1963. An earthquake near Brig (18 km distance, Richter magnitude 5.3, Mercalli intensity VIII) occurred on 23rd March 1960 and most likely triggered the described event.

In conclusion, it can be stated that the present thesis delivered new insights into anatomic tree responses that can serve as a basis for future dendrogeomorphic studies. In the first part, all anatomic growth features following rockfall impacts were assessed and quantified. The present data provide the potential to facilitate and to specify future investigations and reveal the limitation of tree-ring analysis in rockfall research. In the second part, several practical applications were shown. It is possible to reconstruct past frequency, to assess bounce heights and to identify possible rockfall triggers using dendrogeomorphic methods. Data provided by tree-ring analyses offer numerous practical uses such as supporting hazard assessment procedure, identification of local rockfall triggers, provide valuable data for the dimensioning of protections measures, to calibrate and verify rockfall models, or to assess the effect of climate change on rockfall in order to predict future rockfall activity.

♣♣♣♣♣

ZUSAMMENFASSUNG

Geomorphologische Aktivitäten verändern ununterbrochen das äussere Erscheinungsbild alpiner Regionen. Steinschlag ist dabei einer der am häufigsten auftretenden Prozesse, welcher grosse Schäden an Infrastruktur verursachen und zum Verlust von Menschenleben führen kann. Die ständige Ausbreitung des menschlichen Siedlungsraumes in alpine Gebiete sowie eine verstärkte Freizeitgestaltung im Outdoor-Bereich erhöhen fortlaufend das Gefahrenpotential. Ein fundiertes Prozessverständnis ist als Grundlage einer adäquaten Gefahrenbeurteilung unumgänglich und erlaubt es, künftige Schäden durch eine sorgfältige Raumplanung zu vermeiden.

Da Steinschlag ein plötzlich auftretender und selten vorhersagbarer Prozess ist, kann er kaum direkt beobachtet oder erfasst werden. Damit ist die Bestimmung von spezifischen Steinschlagparametern mit erheblichen Schwierigkeiten verbunden. Da sich Echtzeitbeobachtungen von Steinschlagereignissen nicht zur breiten Datenerhebung eignen, bleibt die Rekonstruktion von vergangenen Abgängen als Alternative. Das Auswerten von Archivdaten ist für Steinschalgrekonstruktionen jedoch ungeeignet, da, wenn überhaupt, nur solche Ereignisse niedergeschrieben wurden, welche Bauten zerstört oder Menschenleben gekostet haben. Mittlere bis kleinere Abgänge sowie Ereignisse fernab von bewohnten Gebieten fehlen in den Aufzeichnungen. Sämtliche Methoden, welche bislang zur Rekonstruktion von Steinschlagereignissen verwendet wurden, lieferten höchstens allgemeine und unpräzise Resultate.

Immer fortschrittlichere Steinschlag-Modelle versuchen heutzutage die effektiv gefährdeten Gebiete auszuweisen. Modellberechnungen müssen jedoch stets mit realen Daten abgeglichen und verifiziert werden. Damit kann ein fehlerhaftes Modell erkannt, kalibriert und berichtigt werden. Dieser Abgleich zwischen Modell und Realität bedarf akkurater Daten vergangener Ereignisse. Bis anhin existierte aber keine Methode, welche zuverlässig benötigte Steinschlagdaten rekonstruieren konnte.

Ein Ausweg bietet hier die Dendrogeomorphologie. Es ist die einzige bekannte Methode, welche vergangene Ereignisse jahrgenau lokalisieren, datieren und die eine Vielzahl von Steinschlagparametern (Ursprungsgebiet, Haupttrajektorien, Frequenz, Magnitude, Saisonalität oder Auslösefaktoren) *in situ* akquirieren kann. Ziel dieser Studie war es deshalb, die Methode der Dendrogeomorphologie zu verfeinern und an einigen Fallbeispielen anzuwenden. Damit soll das vorhandene Potential der Dendrogeomorphologie, gerade auch hinsichtlich praktischer Anwendbarkeit, demonstriert werden. Ziel der ersten beiden Studien war es deshalb, die spezifischen anatomischen Auswirkungen eines Steinschlagaufpralles auf einen Baum zu verstehen, um in den beiden weiteren Artikeln Beispiele praxisrelevanter Anwendbarkeiten aufzuzeigen.

In der ersten Studie wurden Stammscheiben von 111 Steinschlagverletzungen in Lärchen (*Larix decidua* Mill.) sowie in Fichten (*Picea abies* (L.) Karst) untersucht, um die tangentiale Ausbreitung anatomischer Baumreaktionen zu quantifizieren. Die Ergebnisse zeigen auf, dass der Baum nach einer Steinschlagverletzung mit der Bildung von Kallusgewebe und vor allem von traumatischen, tangentialen Harzkanalreihen (TRD) reagiert. Lärchen bilden dabei auf 4.1% des nicht zerstörten Gewebes Kallusgewebe und auf 34% TRD. Fichten reagieren mit denselben anatomischen Veränderungen in ähnlicher Ausprägung mit 3.5% Kallusgewebe und 36.2% TRD. Die Entwicklung der Reaktionen über mehrere Jahre zeigt hingegen klare Unterschiede. Währenddem in den Lärchen drei Jahre nach der Verletzung nur noch in 10% der Proben TRD gebildet wurden, können diese in den Fichten fünf Jahre später immer noch in über 50% der Bäume aufgefunden werden. Während der Datenauswertung konnte ausserdem ein saisonaler Unterschied im Auftretenszeitpunkt der TRD festgestellt werden. In Gewebe, welches sich weiter weg von der Verletzung befindet, treten TRD später auf als in unmittelbarer Nachbarschaft. In einem letzten Schritt wurde die Intensität der Reaktion mit verschiedenen Baumparametern verglichen. Dabei zeigte sich, dass die tangentiale Ausbreitung der TRD positiv mit der Breite der Verletzung korreliert, wenn auch nur schwach. Leicht negative Korrelationen gab es mit dem Baumalter und dem Baumdurchmesser. Dies bedeutet, dass junge Bäume mit geringem Durchmesser weniger TRD bilden als alte Bäume mit grösserem Durchmesser.

Der zweite Artikel behandelte ebenfalls die anatomischen Baumreaktionen nach Steinschlagereignissen, jedoch nicht nur in der Ebene der Verletzung, sondern auch in axialer Richtung. Der Fokus wurde in dieser Studie auf die Ausprägung und Auftretenswahrscheinlichkeit von TRD und Druckholz gelegt. Dabei wurden 16 komplette Bäume (Lärchen, Fichten und Weisstannen [*Abies alba* Mill.]) mit insgesamt 54 Verletzungen in Segmente von jeweils 10 cm Höhe zerteilt, woraus 820 Stammscheiben resultierten. Es zeigte sich, dass Fichten tangential und tangential TRD in grösserer Entfernung bilden als Lärchen. Weisstannen hingegen bilden vergleichsweise kaum TRD. Die Position direkt oberhalb der Verletzung weist bei allen Baumarten die grösste Wahrscheinlichkeit auf TRD zu bilden. In Fichten und Lärchen treten an dieser Stelle in über 90% der Fälle TRD auf. Im Allgemeinen kann gesagt werden, dass oberhalb der Verletzungen mehr TRD auftreten als unterhalb. Es konnte keine Korrelation zwischen Grösse oder Fläche der Verletzung und Ausbreitung der TRD festgestellt werden. Auch hier bilden Fichten im Schnitt nach der Verletzung am längsten TRD (3.5 Jahre), gefolgt von Lärchen und Weisstannen (je 1.5 Jahre). Die Jahrringanalyse ergab zudem, dass es zusätzlich zur tangentialen auch eine vertikale zeitliche Verschiebung im Auftreten der TRD gibt. Diese weist keine grossen Unterschiede zwischen den verschiedenen Baumarten auf. Eine weitere Reaktion auf einen Blockaufschlag ist die Ausbildung von Druckholz, mit welchem sich der Baum nach einer Schrägstellung wieder in eine vertikale Position zu bringen versucht. In der Ausprägung konnten signifikante Unterschiede zwischen den Baumarten ausgemacht werden. Währenddem in den Lärchen nur in 8% aller Verletzungen Druckholz auftrat, konnte es in 50% der Fichten und gar in 88% der Weisstannen aufgefunden werden.

Der dritte Artikel sollte einige mögliche Anwendungen der Ergebnisse der ersten beiden Untersuchungen aufzeigen. In einem Steinschlaghang in Saas Balen (Wallis, Schweiz) wurden 154 Wunden von 32 schwer durch Steinschlag beschädigten Bäumen (Lärchen, Fichten, Arven [*Pinus cembra* L.]) untersucht. Die vorhandenen Baumreaktionen wurden verwendet, um den präzisen Auftretenszeitpunkt aller Verletzungen zu bestimmen. Die daraus resultierende Rekonstruktion der Steinschlagfrequenz lässt darauf schliessen, dass es trotz vergangenem Klimawandel zu keiner Aktivitätszunahme in den letzten 50 Jahren gekommen ist. Es zeigte sich, dass 76% aller Verletzungen in der Wachstumspause des Baumes (Anfang Oktober – Ende Mai) auftreten und nur sehr wenige (2.6%) zwischen Mitte Juli und Anfang Oktober. Dies lässt darauf schliessen, dass Frostwechselprozesse im Frühling Hauptauslöser für reguläre Steinschlagabgänge sind. Es konnten jedoch zwei Jahre mit stark erhöhter Aktivität bestimmt werden (1995 sowie 2004), wobei es in beiden Jahren zu aussergewöhnlichen Starkniederschlagereignissen im Spätherbst respektive Frühling gekommen ist. Die Analyse der Verletzungsrichtung liess in einem letzten Schritt Rückschlüsse auf das hauptsächliche Herkunftsgebiet des Steinschlages zu.

Beispiele weiterer Anwendungsmöglichkeiten werden im vierten und letzten Artikel aufgezeigt. Insgesamt wurden für diese Untersuchung 167 Bäume (Lärchen, Fichten und Arven) eines Steinschlaghanges in Saas Balen (Wallis, Schweiz) beprobt. Mit den daraus resultierenden 937 Proben konnten 650 vergangene Steinschlagereignisse datiert werden. Auch hier konnte kein allgemeiner Anstieg der Steinschlagaktivität während der letzten 50 Jahre festgestellt werden.

In einem ersten Schritt wurde rekonstruiert, wann und wie der Hang bewaldet wurde. Die Besiedelung begann in den 1950er Jahren in unterem zentralen Teil und setzte sich dann im Laufe der Jahre immer mehr nach oben hin fort. Die Analyse der Verletzungshöhen ergab, dass die durchschnittliche Sprunghöhe fallender Blöcke im Hang 85 cm beträgt, rund zwei Drittel der Wunden befanden sich unterhalb von einem Meter, die maximale rekonstruierte Sprunghöhe ist 4.5 m. Die grosse Datenmenge erlaubte in einem weiteren Schritt die räumliche Analyse von Sprunghöhe, Steinschlagaktivität sowie der Wachstumsbedingungen. Die grössten Sprunghöhen treten am Rand der Bewaldung auf, die geringsten im zentralen Tel des Untersuchungsgebietes. Dies belegt die Schutzwirkung eines riesigen Blockes oberhalb des Baumbestandes sowie des Waldes selber weiter unten. Die Steinschlagaktivität ist in dem Waldteil am grössten, welcher der hauptsächlich Steinschlag verursachenden Felspartie zugewandt ist. Sie wird jedoch innerhalb des Waldes rasch deutlich geringer, womit die Schutzwirkung des Waldes bestätigt werden konnte. Die Auswertung von Luftbildern dokumentierte schliesslich die Loslösung des grossen Felsblockes direkt oberhalb des Untersuchungsgebietes zwischen den Jahren 1958 und 1963. Wahrscheinlicher Auslöser des Ereinisses war ein Erdbeben der Richterskala 5.3 (Mercalli Intensität VIII), welches am 23. März 1960 in Brig auftrat.

Abschliessend kann gesagt werden, dass

diese Dissertation als Grundlage für künftige dendrogeomorphologische Untersuchungen dienen kann. Im ersten Teil wurden sämtliche anatomische Baumreaktionen nach Steinschlagaufprall erfasst und quantifiziert. Vorliegende Daten erleichtern und präzisieren kommende Untersuchungen, zeigen aber gleichzeitig auch Grenzen der Jahrringanalysen in der Steinschlagforschung auf. Im zweiten angewandten Teil wurde gezeigt, dass mit Hilfe der Dendrogeomorphologie differenzierte Aussagen zu vergangener Aktivität, Frequenz, Sprunghöhen sowie Rückschlüsse auf mögliche Auslösefaktoren getroffen werden können.

Dendrogeomorphologische Daten haben damit eine Vielzahl von praktischen Anwendungsmöglichkeiten, beispielsweise; als Unterstützung bei Gefahrenbeurteilungen, zur Bestimmung von lokalen Auslösefaktoren, als Hilfe bei der Dimensionierung von Schutzbauten, zur Kalibrierung und Verifizierung von Steinschlagmodellen oder um den Effekt der Klimaveränderung auf Steinschlag zu bestimmen um künftige Entwicklungen präziser prognostizieren zu können.

✦✦✦✦✦

Acknowledgements

Here we are, at the very end of my PhD thesis. Even though my name stands on top of the first page, this work would not have been completed without the support of numerous people. Many thanks for accompanying me on my journey to the following persons:

First of all, I would like to express my gratitude to Prof. emeritus Michel Monbaron for giving me the opportunity of accomplishing this PhD thesis and taking care of business even beyond his retirement. Many thanks as well to Prof. Dr. Reynald Delaloye who supported me ever afterwards and gave me the needed space for diligently finishing present thesis.

I am very thankful to Prof. Dr. David Butler and to Dr. Luuk Dorren for their willingness to act as external jury members and for many helpful comments on the manuscript. Thank you very much for making the effort and travelling all the way to Switzerland and finally being present at the PhD defence, my public presentation and the following evening.

Very special thanks go to Dr. Markus Stoffel. First of all, thank you for encouraging me in doing a PhD thesis and for always believing in me. Thank you moreover for your endless support during all the years of planning, the exhausting fieldwork, the tedious laboratory analysis and the scientific writing. Many outcomes of our countless and fruitful discussions can be found on the following pages. However, Markus, most of all I would thank you for your friendship.

Most thanks to my dear companion Dr. Michelle Bollschweiler. Thank you for your endless support wherever and whenever I needed it. Thank you for your effort even in the steepest slopes in the field (I still remember you half frozen but still drudging in that "summer"), for your assistance in the lab, for the critical examination of all my manuscripts, but especially for all the patience and encouragement outside the scientific world. Having you at my side made my journey a pleasure.

During my thesis, I was supported by many other enjoyable people, among them Dr. Oliver Hitz. Thank you for sharing the office with me and for many helpful comments on sample preparation, wood anatomy as well as for many jovial bar excursions.

I would like to address my appreciation to Dr. Moe Myint for his essential introduction into the world of advanced informatics and his endless seemingly kindness.

I address my thanks to the Cemagref institute, namely Dr. Fred Berger, and the technical stuff, Pascal Tardif and Eric Mermin, for cutting down part of my sample trees and for shipping them all the way to Switzerland.

Further thanks go to all the students who assisted me in the field or during sample preparation: Lautaro Correa, Nathalie Abbet and Susanne Widmer.

The thesis at the Department of Geosciences, Geography, University of Fribourg was accomplished in a pleasant atmosphere. I would like to thank all the numerous past and present professors, assistants, students and technical stuff of the institute for many interesting discussions and countless aperitifs.

I would like to address my gratitude to "our" carpenter Michel Torche who installed all the heavy machines and patiently explained their functioning. Sorry for all the blades I accidentally ruined. Thanks to Daniel Cuennet for the sawing and polishing in the carpenter's workshop and to Urs Andenmatten for supporting me in the field and for mending the broken chainsaw.

Working in the Dendrolab.ch was always a pleasure. Besides the people already mentioned there were many others who contributed to this agreeable environment: Patrick Aeby, Estelle Arbellay, Karin Bourqui, Nathalie Chanez, Michael Graupner, Astrid Leutwiler, and bowling world champion Romain Schlaeppy.

Life at University sometimes is not as simple as it seems. Sometimes, there can be lots of paper work: Thank goes to Sylvie Bovel-Yerly, Marie Descloux, and Marianne Zbinden for their support in administration, I still curse that fax machine.

Thank you William Christopher Edward aka Bill Harmer for proof-reading part of unfinished manuscripts.

Writing a thesis is mainly brain work in front of the computer, sports always gave me the needed counterbalance. Many thanks therefore to my always committed soccer team. It's the passion that counts, mates, not the glory!

Last but not least I would like to thank my dear family. During all the years of studying and writing my thesis, I always got the support I asked for, and moreover the support I did not ask for. Never-ending thanks for giving me the freedom of choice, Yvonne and Elmar.

♦♦♦♦♦

Chapter A

General Introduction

1 INTRODUCTION

1.1 BACKGROUND

Rockfall is one of the most common geomorphic processes in mountain regions and has the potential to cause damage to infrastructure or even lead to loss of life (Gardner 1970, Porter and Orombelli 1981, Badger and Lowell 1992, Bunce et al. 1997, Hungr et al. 1999, Erismann and Abele 2001, Sass 2005). Rockfall therefore has become one of the most studied geomorphic processes in the alpine environment.

Swiss regulations on risk and hazard oblige the cantonal authorities (primal administrative unity) to compile hazard maps for the entire territory. If an area is classified as endangered, future development becomes restricted or forbidden by law if no special protection measures are taken (Lateltin 1997, Raetzo et al. 2002). As a consequence, the process of classifying requires careful and integrated assessment of hazards and risks. Federal authorities have thus developed a procedure based on three steps for the analysis of natural hazards (Lateltin 1997, Raetzo et al. 2002). The first step is the identification of the hazard, followed by the assessment of its temporal occurrence and magnitude for determining the hazard potential. Finally, hazard maps have to be produced and, if necessary, risk management strategies, such as the design of protective countermeasures, have to be implemented. To ensure an ideal functioning of any protection measure, specific properties of hazardous processes must be known. In case of rockfall hazards, potential runout zones as a function of source area, boulder parameters and slope properties must be identified. Due to the sudden occurrence and unpredictability of rockfall events, they can rarely be observed or recorded. Therefore, determining specific rockfall properties such as source area, main trajectories, frequency, magnitude, seasonality, or rockfall triggers poses major difficulties. As real-time observations of rockfall are not suitable for data gathering, the reconstruction of historic events remains as alternative data source.

However, there are very limited possibilities for obtaining knowledge of past rockfall events. One possibility of studying past events is the search in archival databases (Dussauge-Peisser et al. 2002, Glade and Lang 2003, Hantz et al. 2003). Available data on rockfall remain fragmentary, as only a minority of past events is documented. Data on major events with loss of life or massive destruction of human infrastructure can probably be found. Small-scale events or events in less populated areas where normally not recognized or if they were, not noted down as they had no significance to the population. Besides direct *in situ* observation (Luckman 1976, Douglas 1980, Gardner 1980, 1983) of rockfall, there exist only very few approaches to reconstruct past events, such as the dating of deposits on the talus slopes using lichenometry (Luckman and Fiske 1995, McCarroll et al. 1998) or decay rates of cosmogenic nuclides (Nishiizumi et al. 1993, Colin et al. 2004, Becker and Davenport 2005). However, all methods do each have several deficits. *In situ* observations are very time consuming and can be undertaken for only a short period of time. Studying rockfall deposits moreover has the very basic limitation of studying the process exclusively in the deposition zone years after the boulder has fallen. No information can be given about the precise year of the event, the source area, trajectories or bounce heights. Reconstruction of rockfall so far allowed for a better understanding of the process and was of no use in hazard analysis and risk assessment. In order to determine the potential risk of a specific area, numerical rockfall models were used. There exists a large variety of different rockfall models to assess the trajectories and outer boundaries of endangered runout zone (Azzoni et al. 1995, Meissl 2001, Guzzetti et al. 2002, Dorren 2003). However, a crucial part for any kind of modeling is the model verification, thus to examine if computed values are consistent with reality. Model computing flawed outcomes can be calibrated in order to minimize prediction errors.

Furthermore, there has been much debate on the existence of an increase of rockfall frequency due to ongoing climate change and its effect on the alpine environment. Different authors project an increase of rockfall rates as a consequence of decreasing slope stability induced by melting processes (Gruber et al. 2004, Krautblatter and Moser 2006). As there is no data available on long-term behavior of rockfall frequencies, it was not possible to determine the effect of global change on rockfall so far.

Until recently, no method existed to reconstruct rockfall parameters and activity. Stoffel et al. (2005a, b) and Perret et al. (2006) used dendrogeomorphic approaches in order to investigate past activity and revealed the high potential of tree-ring data in rockfall research. In contrast to other methods where mainly rockfall deposits were analyzed, dendrogeomorphology gathers data *in situ* where rockfall occurs.

1.2 Aims of the study

The present thesis aims at reconstructing rockfall activity and determines its specific parameters with the study of tree-ring series. The high potential of dendrogeomorphology in rockfall research should be demonstrated by establishing the required fundamentals and by showing numerous possible applications in practice. Obtained data finally should lead to a better understanding of the rockfall process.

The goal of the first part of the study was to elaborate the fundamentals of anatomic tree response to natural rockfall. This study wishes to describe and quantify different types of anatomical tree reactions after rockfall so as to extend and specify conventional dendrogeomorphic methods. Decoding of anatomical tree response following natural impacts allows for a precise dating of tree injuries caused by past events.

The aim of the second part was to use the fundamental anatomical findings in order to demonstrate possible applications. Combining the data of precise event history of many individual trees of a forest stand allows for a creation of a global picture of rockfall behavior on an entire slope. Therewith, past rockfall activity with its spatial behavior was reconstructed on a case study slope in the Valais Alps, Switzerland. Additionally, it was the goal to assess numerous specific characteristics of the rockfall process, such as principal season of activity, frequency and magnitude of events, possible triggers, main source area, and bounce heights across the slope.

There are different practical uses for reconstructed data on rockfall parameters; knowing past rockfall frequencies allows for a investigation of the effect of ongoing climate change on rockfall activity and assists in the prediction of future rockfall rates. Determination of peak activity years and main rockfall season permits identification of local rockfall triggers, and information on bounce heights provide valuable data for rockfall protection measures. Determination of main trajectories, runout zones and bounce heights can help the calibration of existing or future rockfall models and support local hazard assessment.

1.3 STRUCTURE OF THE THESIS

This thesis starts with an overall introduction (Chapter A1), followed by basic information on the rockfall process (Chapter A2) and on its interaction with trees (Chapter A3). In Chapter A4, fundamental methods and applications of dendrogeomorphology are given. The study sites were then localized and presented in Chapter A5, followed by the bibliography of Chapter A (Chapter A6).

Chapter B represents the fundamental part of the thesis and is composed of 2 publications. Chapter B1 and B2 cover anatomical growth responses of different conifer species following rockfall impact. Chapter B1 contains results on the tangential extent of rows of traumatic resin ducts (TRD), their temporal shift with increasing distance from the wound and on the persistence of TRD formation. Results on the tangential and axial arrangement of TRD, the axial shift as well as on the position and intensity of reaction wood can be found in Chapter B2.

Chapter C gives an overview of applications on the basis of the findings presented in Chapter B. Rockfall frequency, intra-seasonal analysis, determination of major events and origin of rockfall was investigated in Chapter C1. Spatial analysis of rockfall activity, bounce heights, afforestation patterns, general growing conditions and geomorphic changes in the slope over five decades are presented in Chapter C2.

The synthesis can be found in Chapter D. Chapter D1 is composed on an overall discussion of the findings with the final conclusions. An outlook on remaining questions as well as possible further research is given in Chapter D2. Chapter D3 finally completes this section with the bibliography of Chapter D.

This work is complemented with appendices (Chapter E), containing a related paper co-written by the author of the present PhD thesis on a rockfall frequency reconstruction over four centuries.

2 Understanding Rockfall

2.1 Introduction

2.1.1 Definitions

Geomorphic processes are constantly shaping mountainous regions. There exist numerous geomorphic mass-movement processes, such as debris flows, landslides, soil creep or rockslides (Selby 1993). However, one of the most widespread among them is rockfall (Gardner 1970, Erismann and Abele 2001, Dussauge-Peisser et al. 2002, Sass 2005).

In the following, several definitions of rockfall are given, illustrating the large variety of rockfall. As can been seen, events can differ in number of active falling rockfall fragments or in the size of moving boulders:

"a fragment of rock detached by sliding, toppling or falling from a vertical or subvertical cliff, before proceeding downslope by bouncing and flying along parabolic trajectories or by rolling on talus or debris slopes" (Varnes 1978);

"a relatively small landslide confined to the removal of individual and superficial rock fragments from cliff faces" (Selby 1993);

"a displacement of a single fragment or several pieces (...) with an episode of free fall during the movement" (Evans and Hungr 1993);

"the free falling of a newly detached rock from a cliff" (Easterbrook 1993);

"a single mass that travels as a freely falling body with little or no interaction with other solids. Movement is normally through the air, although occasional bouncing or rolling may be considered as part of the motion" (Ritter et al. 2002).

As seen, rockfall definitions remain generally rather unspecific. A common element is the free fall. Rock size, however, is not specified in the first place. According to the above definitions, single rockfall fragments can have the size of tiniest pebbles to immense boulders.

2.1.2 Classification

However, it is useful to further subdivide the unspecific meaning of the term rockfall. There exist many different attempts to classify rockfall after volumes of single fragments (Wentworth 1922, Whalley 1984, Berger et al. 2002).

In Switzerland, three main categories of rockfall processes are distinguished. It is distinguished between isolated falling fragments and collective packets of rocks. Collective packets are further subdivided after the volume involved (BUWAL et al. 1997, Gerber 2001).

The falling of single or several isolated blocks is classified after their diameter, illustrations can be found in Figure A2.

- The term **rockfall** (*Steinschlag*) is used to designate falling, bouncing and rolling of fragments with a diameter less than 50 cm. If the diameter of travelling blocks exceeds 50 cm, the term boulder (*Blockschlag*) is used. In common is the sudden and unpredictable detachment of one or several fragments (Fig. A2.1).

Fig. A2.1 A rockfall event of several m^3 on May 31, 2006 on a highway in Gurtnellen caused two fatalities (Uri, Switzerland, Photo courtesy by ASTRA 2009).

- A movement of collective masses is subdivided after their volume. A fragmented portion of a cliff is designated as **large rockfalls** (*Felssturz*) and reaches volumes up to 100'000 m^3 with velocities ranging from 10 to 40 ms^{-1}. During the downslope movement, the collective mass gets more and more fragmented (Fig. A2.2).

Fig. A2.2 A large rockfall of 500 m^3 occurred on January 27, 2009 in Riein (Graubünden, Switzerland) and destroyed the connecting road (Photo courtesy by Tiefbauamt Graubünden).

- **Very large rockfalls** (*Bergsturz*) designate the simultaneous detachment of enormous volumes from one to several millions of cubic meters. The velocities of such catastrophic events exceeds 40 ms^{-1} with considerable long runout distances (Fig. A2.3).

Fig. A2.3 Picture of a very large rockfall event in Randa (Valais, Switzerland). During several batches between April 18 and May 9, 1991, 30 million m^3 of rock were released (Photo courtesy by Air Zermatt).

In the following, the term rockfall is used to designate the falling, bouncing or rolling of isolated fragments with volumes not exceeding a few cubic meters.

2.2 From top to bottom

2.2.1 Promoters and triggers

On each slope, there pre-exist specific morphological, geological or geotechnical properties that determine, if rockfall can occur potentially. In general, it is possible to distinguish rockfall promoters from rockfall triggers (Dorren 2003). Rockfall promoters "prepare" the conditions for possible events, but they do not trigger rockfall, such as rock strength, joint attitude, presence of open cracks, or potential joint infillings (Dorren et al. 2007). After the cliff was "prepared" by the rockfall promoters, the detachment of rocks is then induced by the rockfall triggers.

A large number of different factors have the potential to trigger rockfall. One of the main processes causing the release of boulders are alternating freeze-thaw cycles in the bedrock (Luckman 1976, Douglas 1980, Gardner 1980, Hétu and Gray 2000, Stoffel et al. 2005a, Stoffel and Perret 2006, Kariya et al. 2007, Matsuoka 2008). Alternating temperatures around the freezing point alone lead only rarely to the detachments of rockfall fragments. As stated by many authors, the additional presence of moisture severely amplifies the effect of freeze-thaw cycles. In combination, these factors act as main triggers for "regular" rockfall events (Matsuoka 1990, 2001, Matsuoka et al. 1997, Sass 1998, Matsuoka and Sakai 1999, Rovera et al. 1999, Braathen et al. 2004, Hall 2007). If temperatures rise, ice loses its stiffness and strength. This process leads to instabilities even if the ice is not melted (Davies et al. 2001). In case of a further temperature increase, ice finally starts to melt and water seeps into cracks. This leads to a major loss of bonding and to higher water pressure in the joints, both resulting in further instability

increase (Dramis et al. 1995, Matsuoka ans Sakai 1999, Davies 2001). If temperature then decreases, joint water begins to freeze, leading to a volume increase. This procedure causes crack widening and controls to a large degree the magnitude of frost action (Walder and Hallet 1986, Fahey and Lefebvre 1988, Matsuoka et al. 1997, Goudie and Viles 1999, Matsuoka 2001, Ishikawa et al. 2004). Matsuoka (2008) distinguishes three scales of crack opening: small opening by short-term frost cycles, slightly larger movements during seasonal freezing and occasional

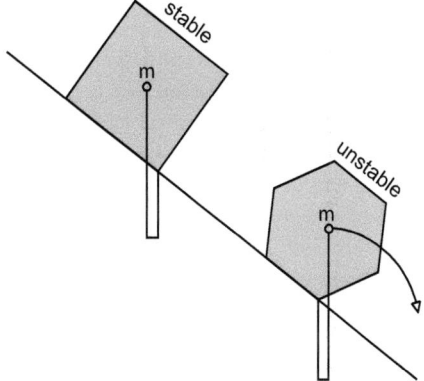

Fig. A2.4 Illustration of the influence of rockfall fragments on their stability. The square shaped boulder remains in position, whereas the hexagonal shaped fragment starts travelling (modified after Descoeudres 1990).

large opening originating from refreezing of snow-melt water during seasonal thawing. This freeze-thaw cycle is very complex as rock temperature depends on many local factors, such as weather conditions, slope angle or slope orientation (Rovera et al. 1999, Matsuoka et al. 1997, Hall 2007, Matsuoka 2008). The loss of ice bondage can also be caused by a general uplift of permafrost, induced by global warming (Haeberli and Beniston 1998, Haeberli 1999, Schiermeier 2003, Gruber et al. 2004, Krautblatter and

Moser 2006). Resulting cohesion lost would then lead to increased rockfall activity (Sass 2005).

However, water has the potential to trigger rockfall events outside the freeze-thaw cycle as well. Higher temperatures in summer cause durable melting of ice and releases large amounts of water. Melted water then fills cracks and joints and increases the pore pressure. Rockfall in the warmer period of the year can therefore be triggered by water without passing through short-term freeze-thaw cycles (Penck 1924, Rapp 1960, Gardner 1970, Luckman 1976, Haeberli 1996). Matsuoka and Sakai (1999) and Dramis et al. (1995) for instance recorded most rockfall events during summer season in the Japanese and Italian Alps respectively.

Water-induced rockfall triggering can also be caused by meteorological events (Sandersen et al. 1997), most of all by heavy rainfall (Bjerrum and Jorstad 1966, Luckman 1976, Peckover and Kerr 1977, Kotarba and Strömquist 1984, Butler 1990, Sandersen et al. 1996, Sass 1998, Rosser et al. 2005). Of course, freeze-thaw cycles, melting processes and heavy rainfalls cannot be seen as completely independent factors, they often act mutually. Several authors report on the combination of triggering factors as rainfall and melt water (Rapp 1960, Gardner 1980, Nyberg 1991), as an interaction of rainstorm and freeze-thaw processes (Nyberg 1991, Krautblatter 2003, Decaulne and Sæmundsson 2006) or the combination of all three processes (Gruner 2008).

Several studies report on rockfall events triggered by earthquakes (Case 1988, Bull and Brandon 1998, Marzoratti et al. 2002, Braathen et al. 2004, Sepulveda et al. 2004, Becker and Davenport 2005). Following Bull and Brandon (1998), an earthquake with a Richter magnitude 7 can induce rockfall at a distance up to 400 km. Keefer (1984, 2002) gives a minimum magnitude of 4.0 that is required to cause rockfall. However, Rodríguez et al. (1999) report on an event already triggered by a magnitude 2.9 earthquake.

Finally, there exist few other processes that have the potential to cause rockfall, such as chemical weathering (Whalley 1974) or wind (Blackwelder 1942, Rapp 1960). However, rockfall triggered by these processes remains scarce.

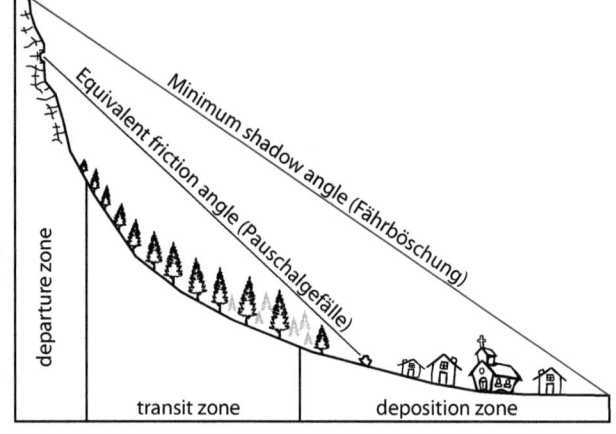

Fig. A2.5 Subdivision of a rockfall slope into departure, transit and deposition zone with graphical representation of the "minimum shadow angle" and the "equivalent friction angle" (modified after Heim 1932, Körner 1980, Gerber 1998, Kienholz 1998). See text for further explanation.

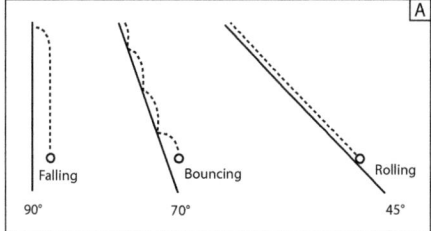
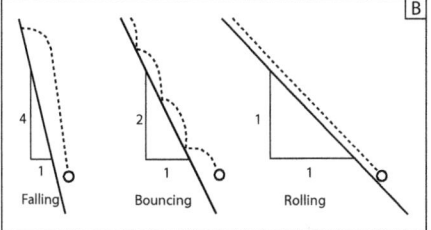

Fig. A2.6 Thresholds for falling, bouncing and rolling following Dorren (a, 2003, adapted) and Schweigl et al. (b, 2003, adapted).

2.2.2 ZONES OF A ROCKFALL SLOPE

A rockfall slope can be divided into three parts: the departure zone, the transit zone and the deposition zone (Fig. A2.4). To release rocks and boulders, the departure zone requires a minimum slope angle of 30° (John and Spang 1979, Jahn 1988, Gsteiger 1993, Gerber 1998, Schwitter 1998). As can be seen in Figure A2.4, the critical slope angle at which fragments start travelling also depends on the shape of the fragments.

It can be distinguished between primary and secondary rockfall (Luckman 1976). The initial detachment of a rock fragment from the cliff is designated as primal rockfall. The concept of secondary rockfall describes the situation where already stopped rockfall particles remain in intermediate storages until they are newly removed (Krautblatter and Dikau 2007). Secondary rockfall can mainly be observed during heavy precipitation events (Beylich and Dandberg 2005, Krautblatter and Dikau 2007).

After release in the departure zone, rocks start traveling downslope and enter the transit zone. This is the middle part of the slope where no primary rockfall occurs, while slope angles remain too steep to permanently stop moving fragments. Traveling boulders in the transit zone are mainly influenced by the underground (surface roughness, dampening) or by contact with obstacles (trees, large boulders). Boulders in the transit zone may be stopped intermediately but can then be reactivated and travel further down as secondary rockfall. However, rockfall gen-

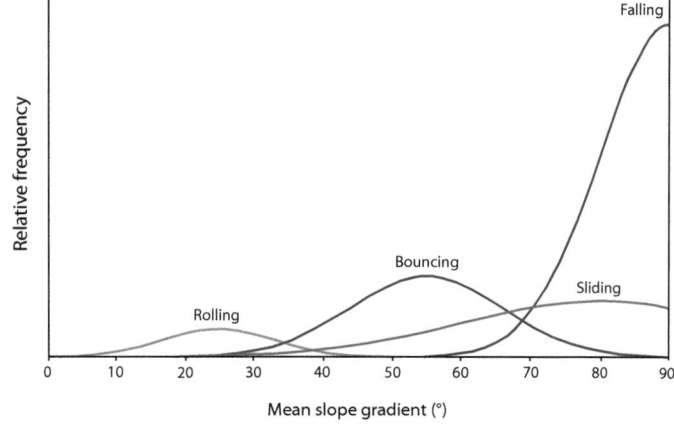

Fig. A2.7 Relative frequency of the different motion types, depending on the slope gradient (Dorren 2008).

Fig. A2.8 During ground contact, a large proportion of energy is lost. Example of a rockfall crater following ground contact in Balmatten (Valais, Switzerland). See the pencil (white arrow) for scale.

Fig. A2.9 Example of a discoid shaped rockfall fragment that has the potential to tilt up and reach extraordinary velocities and runout distances. Picture from Täschgufer (Valais, Switzerland).

Fig. A2.10 Injured tree in Saas Balen, Valais, Switzerland, witnessing of remarkable rebound heights of falling rockfall fragments (approx. 3 m, white arrow).

erally passes through this intermediate transit zone.

If slope gradients drop below 30°, rockfall fragments generally come to a stop in the deposition zone (BUWAL et al. 1997). This area of accumulated rockfall fragments at the foot of a cliff or slope is called scree slope or talus slope (Allaby and Allaby 1990). However, a deposit area can also act as a secondary rockfall source area (Dorren et al. 2007).

Evans and Hungr (1993) proposed the notion "minimum shadow angle" in order to describe the general aspect of rockfall slopes (Fig. A2.5). Thereby, the highest point of the talus slope and the farthest deposition point were considered (equivalent to the *Fahrböschung*, suggested by Heim (1932)). In contrast, the "equivalent friction angle" (*Pauschalgefälle*) describes the angle formed by a line connecting the center of gravity of the departing mass with that of the deposited mass (Kienholz 1998).

2.2.3 ROCKS MOVING

As like for all mass-movement processes, the driving force behind rockfall is gravitation. Physically speaking, potential energy is transformed into kinetic and rotational energy. Three main modes of motion can be distinguished: falling, bouncing and rolling. The motion type is mainly influenced by the slope angle (Fig. A2.6).

In general, travelling starts with a short period of sliding (Azzoni et al. 1995). As soon as the fragment starts moving, friction forces occur, depending on boulder shapes and surface properties. With increasing friction forces, sliding passes over to rolling

2. Understanding Rockfall - 25

Fig. A2.11 A house severely damaged by rockfall in the Gottéron Valley, Fribourg, Switzerland after a rockfall event on March 6, 2005 (Photo courtesy by Vincent Menoud and Bertrand Yerly).

Fig. A2.12 Car hit by a falling boulder in Saas Balen, Valais, Switzerland on January 19, 2007 (Photo courtesy by Kantonspolizei Wallis).

Fig. A2.13 Rockfall net as a possible but expensive protection measure (Balmatten, Valais, Switzerland).

(Erismann and Abele 2001). Of course, pure rolling rarely exists in nature, as perfectly spherical rock fragments and very plane surfaces would be required. However, it can be stated, that rolling occurs if the irregularities in the shape of travelling rocks are larger compared to the ones on the slope surface (Azzoni et al. 1995). Theoretical rolling velocities can be calculated exclusively as a function of slope angle and surface roughness (Meissl 1998, Woltjer et al. 2008).

During rolling, a rockfall fragment starts accumulating rotational energy. In the literature, different values for the proportion of the rotational energy on the overall energy can be found, given values range from 10% (Jaboyedoff et al. 2003), to 20% (Gerber 1994) and 30% (Jahn 1988, Chau et al. 2002).

As demonstrated, irregularities in shape and surface transform the rolling into bouncing, especially at higher velocities (Erismann and Abele 2001). Following Woltjer et al. (2008), a boulder starts jumping from the moment when rotational velocity exceeds the translational velocity. Gerber (1998) and Schwitter (1998) suggest a minimum slope angle of 35° to pass from rolling to bouncing. Dorren (2003) and Schweigl et al. (2003) however propose an inclination of 45°. After Azzoni et al. (1995), bouncing can be expected if irregularities in boulder shape where smaller compared to the ones of the slope surface.

A steeper slope angle or the presence of a sudden cliff transforms bouncing or rolling into free falling. Again, different critical slope angles that allow falling can be found in literature. According to Ritchie (1963), a minimum slope of at least 76° is required. Dorren (2003) suggests 70°, whereas Erismann and Abele (2001) propose 45-55°. Other authors define threshold values by comparing vertical distances to horizontal distances. Following Schweigl et al. (2003), free falling occurs as soon as the vertical distance travelled by the rockfall fragments exceeds the horizontal distance by a factor of four. Another interesting approach is presented by Dorren (2009), who gives the distribution of relative values for all motion types, depending on the steepness of the slope (Fig. A2.7).

During the free falling, air friction can be neglected (Bozzolo and Pamini 1986a, Erismann and Abele 2001). Surprisingly, free falling seems not to be more effective as rolling, as consequence of substantial energy losses during impacts with the surface (Zinggerler 1989). The effective proportion of energy lost during ground contact depends on the rock consistence, the slope angle, and surface properties (Krummenacher 1995, Erismann and Abele 2001). During the first impact after falling, 75-86% of accumulated energy is absorbed at once (Broilli 1974, Evans and Hungr 1993, Dorren 2003, Fig A2.8). Loose material, soft or wet soils lead to increased energy dissipation during ground contact (Gsteiger 1993, Gerber 1994, Krummenacher and Keusen 1996).

The boulders finally come to a stop in the deposition zone. Due to their higher masses, big boulders accumulate more kinetic energy during travelling downslope, resulting in longer runout distances (Jahn 1988, Erismann and Abele 2001). More spherically shaped rocks tend to have longer runout distances than non-spherical ones. Azzoni and Rossi (1991) describe single cases of discoid fragments that tilted up and behaved like wheels rolling down the slope, thus reaching the longest runout distances (Fig A2.9). The described special case however occurs scarcely (Erismann and Abele 2001). Another rare event is the simultaneous falling of many rockfall fragments, single ones then have the potential to travel far, as they can gather energy from collisions with other rocks, leading to higher velocities (Okura 2000a). Stopping in general occurs through a rather abrupt than a gradual process (Dorren 2003). According to Krummenacher (1995), boulders stop most abruptly if the irregularities on the slope surface are of the same magnitude as the fragment sizes.

2.2.4 ROCKFALL VELOCITIES AND REBOUND PARAMETERS

Velocities of rockfall fragments can range from a few to tens of ms^{-1}. Swiss authorities assume maximum rockfall velocities are reached at 30 ms^{-1} (BUWAL 1997). Azzoni and Rossi (1991) conducted *in situ* tests on two different rockfall slopes and measured average velocities of 10 to 12 ms^{-1} and maximum speeds of 20 ms^{-1}. Rickli et al. (2004) gathered field data from impact craters caused by large rockfall boulders (10 t) and reconstructed jump parabola and travelling velocities. He suggests maximum velocities of 25 ms^{-1} and states that the mass of rockfall fragments influences rebound distances and rebound heights. Rocks with masses of 25 kg reached jump distances of about 2.2 m, whereas large boulders of 10 t were able to jump over distances of 30 m. Dorren et al. (2006a, b) likewise released boulders on a non-vegetated slope and measured average velocities of 11 ms^{-1} and maxima of 30.6 ms^{-1}. The rebound heights averaged 1.5 m with a maximum of 8 m (Fig A2.10). As doubling of velocities leads to four times more energy, it is evident that big boulders reach much higher energy values than smaller ones (Kirkby and Statham 1975, Erismann and Abele 2001, Meissl 2001, Dorren 2003). Rickli et al. (2004) calculated the accumulated energies of the small (25 kg) and the big boulders (10 t) and found differences up to a factor of 6000 (0.5 kJ vs. 3000 kJ).

2.3 Hazard analysis, risk assessment

Rockfall belongs to the most common and destructive mass movements in mountain regions and has the potential to cause damage to infrastructure or even to loss of life (Gardner 1970, Porter and Orombelli 1981, Badger and Lowell 1992, Bunce et al. 1997, Evans 1997, Hungr et al. 1999, Guzzetti 2000, Erismann and Abele 2001, Sass 2005,). Increasing anthropogenic activities in marginal regions results in the construction of new infrastructure and settlements in exposed areas (Fig. A2.11, A2.12). Therefore, an accurate risk assessment becomes more and more inevitable in order to avoid future accidents. Hazard assessment should identify: areas were rockfall occurs, frequency and magnitude of events, and local rockfall triggers. Swiss regulations on risk and hazard oblige the authorities to compile hazard maps for their entire territory. All areas where categorized in different hazard classes in order to decide, whether future development becomes restricted or not (Lateltin 1997, Raetzo et al. 2002). Federal authorities apply a three step procedure that is based on the analysis of natural hazards (Raetzo et al. 2002). Hazard Identification is the first step in this procedure, followed by the assessment of frequency and magnitude in order to determine the hazard potential. In a last step, the hazard maps need be produced and, if necessary, risk management strategies have to be implemented A2.13).

Different approaches can be used to assess rockfall hazard, such as Slope Mass Rating Systems (Romana 1988), Rockfall Hazard Rating Systems (Pierson et al. 1990, adapted by Budetta 2004) and Matterrock (Rouiller et al. 1997). All these methods evaluate different parameters (geology, rockfall history, rockfall volume or triggering factors) to produce an overall hazard rating. Therefore, knowledge on all different aspects is required. However, these applications do not result in specific rockfall rates. So far, there exist only few studies that determined and quantified rockfall hazard. Evans and Hungr (1993) determined the "landing probability" of boulders in the shadow of talus slopes. Rockfall hazard and risk assessment in the Yosemite Valley, California, USA were investigated by Guzzetti et al. (2003). Other studies have mainly been conducted along transportation corridors. A risk analysis methodology that determines the probability of loss of life on British Columbia Highway was developed by Bunce et al. (1997). Other studies on highways were conducted by Hungr and Beckie (1998), Hungr et al. (1999), Budetta (2004) or Maerz et al. (2005). However, results of these studies mostly remain unspecific and do not give quantitative data for specific slopes. Corominas et al. (2005) were able to quantify the expected annual loss of life on a slope in Andorra and determined protection effect of rockfall protection fences. To do so, detailed information on geology, historic events, meteorological conditions, and vulnerability values of infrastructures as well as the use of dendrogeomorphic methods were necessary. Stoffel et al. (2005b) and Perret et al. (2006) finally determined specific values of mean yearly rockfall events per surface exclusively using dendrogeomorphic methods.

2.4 Modeling

To assess rockfall hazards, a profound knowledge of numerous field properties is required. Careful hazard assessment on a local scale is very time consuming and of great cost. Even an appropriate evaluation of all parameters results at most in a good estimation of hazarded zones and remains a subjective appraisal. Therefore, there was a trend in recent years towards rockfall modeling. The following structure and presented literature is based on Guzzetti et al. (2002), Dorren (2003), and Dorren et al. (2007).

2.4.1 EMPIRICAL ROCKFALL MODELS

Empirical models are based on relationships between topographical factors and length of the resulting runout zone. Numerous case studies were compared in order to find correlations between topographic properties and horizontal travelling distances. Therefore, these models can be referred to as statistical models (Keylock and Domaas 1999). Factors such as maximum vertical drop, volume of rockfall, *Fahrböschung*, minimum shadow angle, energy of a rockfall event, horizontal length of the talus slope, or horizontal length of the free rock face were taken into account to calculate the length of the runout zone. Such empirical models were presented by Tianchi (1983), Toppe (1987), and Evans and Hungr (1993).

2.4.2 PROCESS-BASED ROCKFALL MODELS

Process-based rockfall models describe or simulate the different modes of motion of fragments over the slope surface. In the following, an overview of a vast number of existing rockfall models is given. Most of the available rockfall models are two-dimensional and restricted to the motion of rocks within a vertical plane, lateral movement is excluded (Kirkby and Statham 1975, Piteau and Clayton 1976, Bassato et al. 1985, Falcetta 1985, Wu 1985, Bozzolo and Pamini 1986a, b, Hoek 1987, Bozzolo et al. 1988, Hungr and Evans 1988, Pfeiffer and Bowen 1989, Kobayashi et al. 1990, Pfeiffer et al. 1991, Budetta and Santo 1994, Chen et al. 1994, Azzoni et al. 1995, Chau et al. 1998, Stevens 1998, Keylock and Domaas 1999, Paronuzzi and Artini 1999, and Jones et al. 2000). Nevertheless, lateral movement in rockfall modeling is of importance, in particular if the slope is of concave (along channels) or convex (on the deposition fan) form (Crosta and Agliardi. 2004). However, there still exist only few three-dimensional model approaches (Descoeudres and Zimmermann 1987, Van Dijke and Van Westen 1990, Scioldo 1991, Gascuel et al. 1998, Frattini et al. 2008). Three-dimensional models that simulate changes of kinetic energy of particles as a result of inelastic and frictional collisions with each other and with the slope surface were presented by (Donzé et al. 1999, Okura et al. 2000a, b, Koo and Chern 1998)

2.4.3 GIS-BASED ROCKFALL MODELS

GIS-based models are models that either run in a GIS environment or their raster basis was provided by GIS analysis. To run GIS-based rockfall models, three steps need to be worked through. In a first step, the source area needs to be identified. Then, the fall tracks are determined before the length of the runout zone finally can be calculated (Van Dijke and Van Westen 1990, Meissl 2001, Menéndez Duarte and Marquínez 2002, Ayala-Carcedo et al. 2003, Baillifard et al. 2003, Dorren and Seijmonsbergen 2003, Marquínez et al. 2003, Chau et al. 2004). GIS-based models can also integrate process-based models for trajectory calculations. These types of more realistic models evolved strongly in recent years (Krummenacher and Keusen 1996, Agliardi et al. 2001, Guzzetti et al. 2002, Agliardi and

Crosta 2003, Crosta & Agliardi 2003, Crosta et al. 2004, Woltjer et al. 2008). Few rockfall simulation models take additionally into account the mitigating effect of the existing forest cover (Zinggeler 1989, Liniger 2000, Dorren and Seijmonsbergen 2003, Le Hir 2005, Dorren et al. 2004a, 2006a, Perret et al. 2004, 2006, Kühne 2005, Stoffel et al. 2006a).

However, to obtain reliable data through GIS-based modeling, a detailed data basis is required, including position of source area, potential fragment sizes, or slope properties (angle, inclination, roughness, and obstacles). The reliability of the output can thus only be as good as the quality of the input data. A crucial part for any kind of modeling is the model verification so as to examine if computed values are consistent with the ones occurring in reality. In case of rockfall modeling, this essential step causes some difficulties, as artificial release of geomorphic events *in situ* is not possible in general. Alternatively, past events can be investigated and compared to the values provided by the modeling. To do so, past rockfall frequencies, rebound heights, and trajectories need to be reconstructed.

2.5 ROCKFALL RECONSTRUCTIONS

To establish a reliable hazard assessment and risk analysis, an extensive database including past rockfall frequencies and magnitudes would provide indispensable support. There exist only limited methods in order to obtain knowledge on rockfall events.

Direct observation is probably the most accurate way of studying rockfall properties. However, *in situ* observations are very time consuming and can only be applied for a short period of time on a very limited study site (Luckman 1976, Douglas 1980, Gardner 1980, 1983, Matsuoka and Sakai 1999). As natural rockfall occurs suddenly and unpredictably, there is no guarantee of catching an event at all. Reconstruction of past events is therefore the only possibility to obtain broad knowledge on general rockfall behavior. Studying archival data on rockfall is the most obvious method (Rapp 1960, Abele 1971, Bunce et al. 1997, Hungr et al. 1999, Dussauge-Peisser et al. 2002, Glade and Lang 2003, Hantz et al. 2003, Guzzetti et al. 2004). However, only a fraction of past events are normally recorded. In general, there exist only data on major events that caused loss of life or destroyed human infrastructure. Rockfall in less populated areas is often not recognized or not recorded in archives, as it has no relevance for the population. In recent years, lichenometry was repeatedly used to date rockfall deposits, and, sizes of lichens were compared to yearly growth rates (André 1986, Luckman and Fiske 1995, Bull and Brandon 1998, McCarroll et al. 1998, 2001, Bajgier-Kowalska 2002, Matthews and Shakesby 2004). Finally, decay and accumulation of different cosmogenic nuclides can be used to date deposits, such as ^{230}Th-^{234}U-^{238}U (Ludwig et al. 1992), ^{10}Be-^{26}Al (Nishiizumi 1993 et al.-, Colin et al. 2004) or ^{14}C (Becker and Davenport 2005). Both methods, lichenometry and cosmogenic nuclide dating, present several deficiencies. Obtained dating is not very accurate and only possible within a certain range, the older the deposits, the wider the range. Even the assessment of single decay ratios requires expensive analyses and the availability of sophisticated laboratory equipment. As exclusively deposits are investigated, there exist several serious limitations: No information can be given about the precise year or season of an event, the source area, main rockfall trajectories or occurring bounce heights.

Very recently, dendrogeomorphic methods were applied to assess past rockfall frequencies. Stoffel et al. (2005b) and Perret et al. (2006) demonstrated the high potential of tree-ring analysis by reconstructing up to four centuries of rockfall activity and assessing its seasonality, spatial pattern and recurrence interval.

♦♦♦♦♦

3 ROCKFALL-FOREST INTERACTIONS

3.1 INFLUENCE OF TREES ON ROCKFALL

Forest stands in mountain regions often have an indispensable protection function against natural hazards such as rockfall, landslides, debris flows, snow avalanches, floods, or erosion (Omura and Marumo 1988, Hamilton 1992, Brang et al. 2001, 2006, Margreth 2004, Rickli et al. 2001, 2004). In the case of rockfall, forests have a high potential to lead to an early stop of fragments (Meissl 1998, Perret et al. 2004, Dorren and Berger 2006a, b, Stoffel et al. 2006a). Some illustrations of boulders and rocks stopped by trees are given in Figures A3.1, A32, and A3.3. However, the effect of trees on rockfall depends on their position on the slope.

3.1.1 DEPARTURE ZONE

In the departure zone, trees can both reduce and increase rockfall activity. Their roots retain loose material and prevent it from being detached, resulting in fewer events. How-

Fig. A3.1 Exceptional example of a large boulder of 10 m³ that was stopped by a tree.

Fig. A3.2 A rock of 0.5 m³ came to a stop between two trees.

Fig. A3.3 A fragment with the size of a tennis ball is stuck in the stem after impact.

ever, in order to increase stability, tree roots produce organic acids that attack the rock by chemical weathering (Jahn 1988). They create or widen pre-existing rock joints leading to more fractures in cliffs and increased frost action (Corominas et al. 2005, Frehner et al. 2005). Roots finally have to absorb any movements of the crown induced by wind or snow forces, leading to movements of the entire root-soil plate. Resulting wedge effects lead to increased rock fractioning and to the detachment of fragments (Gerber 1998). The mechanics of energy dissipation from stems to the root-soil plate were recently studied by Lundström et al. (2007a, b). In general, the negative effects of trees in the source areas exceed the positive effects (Dorren et al. 2007). Of course, adult trees in the source area do already act as obstacles and decelerate falling rocks as described in the following section. However, the presence of regular tree stands in source areas is rather unusual.

3.1.2 TRANSITION AND DEPOSITION ZONE

In the transition zone, trees can be seen as obstacles, leading to collisions with falling fragments. Impacts with trees reduce the velocity of moving boulders, leading to a reduction of bounce heights and to shorter runout distances (Brang 2001, Perret et al. 2004, Dorren et al. 2006, Stoffel et al. 2006a). The source area does not need to be extended to allow high rockfall velocities, as maximum travel speed is reached already after 40 m (Gsteiger 1993). There exist only few quantitative data on real-size experiments on forested rockfall slopes. Dorren et al. (2006b) assessed different rockfall parameters on a forested and a non-forested zone at a rockfall slope in Vaujany, France. As can be seen in Table A3.1, travelling velocities and rebound heights (both means and maxima) were considerably reduced by the forest. A slight negative effect of a rockfall protection forest is the wider run-out area with a lateral deviation of 10° from the vertical fall-line to both sides (Jahn 1988, Dorren et al. 2005)

It is known that there is an exponential relationship between diameter at breast height (DBH) and energy dissipation, and that the number of impacts against trees seems to be more important than the efficacy of a single impact (Dorren et al. 2005). However, in a forest, there exists a natural trade-off between tree density and large DBH. It is not possible to grow a dense forest with exclusively large trees (Frehner et al. 2005). Therefore, a compromise has to be found between a forest that causes maximum number of impacts and maximum energy dissipation per impact. Following Ott et al. (1997), Kräuchi et al. (2000), Dorren et al. (2004b), and O'Hara

Rockfall parameters	Non-forested (n = 100)	Forested (n = 102)
Average translation velocity (ms^{-1})	11	8
Average maximum translation velocity (ms^{-1})	15.4	11.7
Maximum translation velocity (ms^{-1})	30.6	24.2
Number of rocks stopped after 223.5 m (end of forested zone)	5	65
Mean rebound height (m)	1.5	1
Maximum rebound height (m)	8	2

Table A3.1 Comparison of traveling velocities and rebound heights between a non-forested and a forested slope (Dorren et al. 2006b).

Zone	Potential contribution of forest	Minimum standard	Ideal standard
Release zone	Medium	'Backbone' trees / No unstable heavy trees	
Transit zone	High / Rocks up to 0.05 m³ (diameter about 40 cm)	Horizontal structure ≥ 400 trees/ha with DBH >12 cm	Horizontal structure ≥ 600 trees/ha with DBH >12 cm
		potentially also coppice	
		Vertical structure / Target diameter[a] appropriate	
	Rocks 0.05 to 0.20 m³ (diameter about 40 to 60 cm)	Horizontal structure ≥ 300 trees/ha with DBH >24 cm	Horizontal structure ≥400 trees/ha with DBH >24 cm
		Vertical structure / Target diameter[a] appropriate	
	Rocks 0.20 to 5.00 m³ (diameter about 60 to 180 cm)	Horizontal structure ≥150 trees/ha with DBH >36 cm	Horizontal structure ≥200 trees/ha with DBH >36 cm
	Additionally for all rock sizes	Horizontal structure / If gaps[b] exist in slope direction: stem distance <20 m / Lying logs and high stumps as complement to standing trees, if no risk of fall	
		Stand meets criteria of minimum site-specific standard	Stand meets criteria of ideal site-specific standard
Runout and deposition zones	High / The effective minimum diameter of trees is considerably smaller than in the transit zone, and lying logs are always effective	Horizontal structure ≥400 trees/ha with DBH >12 cm	Horizontal structure ≥600 trees/ha with DBH >12 cm
		Horizontal structure / If gaps[b] exist in slope direction: stem distance <20 m, potentially also coppice	
		Vertical structure / Target diameter[a] appropriate / Lying logs and high stumps as complement to standing trees	
		Stand meets criteria of minimum site-specific standard	Stand meets criteria of ideal site-specific standard

Table A3.2 Structure of an ideal rockfall protection forest (Frehner et al. 2005, Annex 1, p. 14). [a] The target diameter has to be chosen so that the required stem density with stems of the effective minimum diameter can be permanently maintained. [b] Gap: opening from crown edge to crown edge in pole and old timber stands.

(2006), a mixed multilayered forest is best suitable for sustainable rockfall protection. The specific distribution of DBH depends on boulder sizes, in case of small fragments, the forest should tend toward higher density, in case of large fragments towards larger DBH (Jahn 1988, Stokes et al. 2005, Wehrli et al. 2006, Dorren et al. 2006a). After Frehner et al. (2005), increasingly larger DBH are required (i) the steeper the slope is, (ii) the rounder the fragments are, (iii) the weaker the soil dampening is, (iv) the lower the surface roughness is and (v) the more fragile the trees are. Schwitter (1998) states that the mean DBH of trees in a protection forest should be approximately one third of the size of the falling fragments. This value has been confirmed by Dorren et al. (2005). Frehner et al. (2005) presented detailed guidelines on tree density, DBH, and vertical structure of a rockfall protection forest as a function of specific boulder sizes in each of the release, the transit and the deposition zone (Table A3.2). Many authors describe the best suitable composition of protection forests as well as the methods to apply in order to achieve and maintain its purpose (Chauvin et al. 1994, Wasser and Frehner 1996, Brang et al. 2000, 2006, Motta and Haudemand 2000, Brang 2001, Bebi et al. 2001, Dorren et al. 2004b, Thormann and Schwitter 2004, Brauner et al. 2005, Frehner et al. 2005, Wehrli et al. 2006).

3.2 INFLUENCE OF ROCKFALL ON TREES

A single tree dissipates energy during a rock impact in different ways. The mechanical shock causes local penetration of the fragment at the impact location, deformation and oscillation of the tree stem, as well as rotation and translation of the entire root system. Interestingly, approximately half of the energy released during a rockfall impact seems to be absorbed by the root system (Foetzki et al. 2004, Brauner et al. 2005, Kalberer et al. 2007). More than a third of total energy dissipates in stem bending and stem displacement (Foetzki et al. 2004). Lundström et al. (2009) however conducted trolley tests to simulate rockfall impacts and investigated resulting energy dissipation. They concluded that 75% of released energy was absorbed by the stem and crown, while the remaining 25% were dissipated by the root-soil system. If uprooting is prevented by good anchorage of the tree and if there is no stem breaking, the impact energy is transferred through the stem to the tree crown. This transfer induces a sinusoidal shockwave ("hula-hoop" effect, Dorren et al. 2006a) that often leads to a breaking of the tree top (Fig. A3.4). In general, broad-leaved trees are more resistant to mechanical impact that coniferous trees (Couvreur 1982, Rupé 1991, Peltola et al. 2000, Stokes et al. 2005, Dorren and Berger 2006).

There exist few quantitative data on energy release of falling rocks and energy dissipation capability of trees. Even more, laboratory tests seem not to deliver reliable data for trees growing under natural conditions (Fig. A3.5, Fig. A3.6, and Fig. A3.7). Tests on *Fagus sylvatica* L. (European Beech) revealed twice as high energy dissipation values for trees in the field compared to wood samples in the laboratory (Couvreur 1982). Similar results were obtained by Dorren and Berger (2006), who compared maximum energy dissipation of *Picea abies* (45 cm diameter) obtained by different methods. Highest values of tree resistivity were delivered by real-size rockfall experiments (230 kJ), followed by bending energy (184 kJ), static winching experiments (125 kJ), and standardized dynamic impact tests (8 kJ). As seen, trees are able to dissipate energy in their entire system (the root-soil plate, the stem, and the crown), explaining the enormous differences of dissipation values between laboratory tests and experiments under natural conditions. More detailed data on energy values dissipation of different trees can be found in Dorren and Berger (2006).

3. Rockfall-forest interactions - 35

Fig. A3.4 Impacts often lead to apex loss.

Fig. A3.5 A collision resulted in stem inclination and splitting of the stem.

Fig. A3.6 A tree broken into several pieces after being hit by a fragment.

Fig. A3.7 Even though the individual was destroyed by the impact, the boulder could be stopped.

♣♣♣♣♣

4 Dendrogeomorphology

4.1 Background

During spring and early summer, conifer trees form large tracheids with thin cell walls (earlywood cells). In late summer and early autumn, in contrast, smaller tracheids with thicker cell walls are formed (latewood cells). The smaller lumen of the cells leads to a darker appearance of the latewood. During dormancy (late autumn to early spring), there is no formation of tracheids at all (Schweingruber 1983, 1996, 2001). A sequence of brighter earlywood cells and following latewood cells represents one growth ring (tree ring, annual ring).

The appearance of the ring is influenced by many different internal as well as external factors. Tree species, genetics or age are influencing growth from the inside. A plethora of external factors control tree growth, such as light, temperature, water availability, nutrients disposability, wind, concurrence, or pests (Schweingruber 1996). Therewith, trees constantly record numerous information of their environment. Individuals that grow in proximity are influenced by the same environmental impacts, resulting in similar tree-ring series. Therewith, tree-ring series of unknown age can be dated by juxtaposing them to known sequences. This method

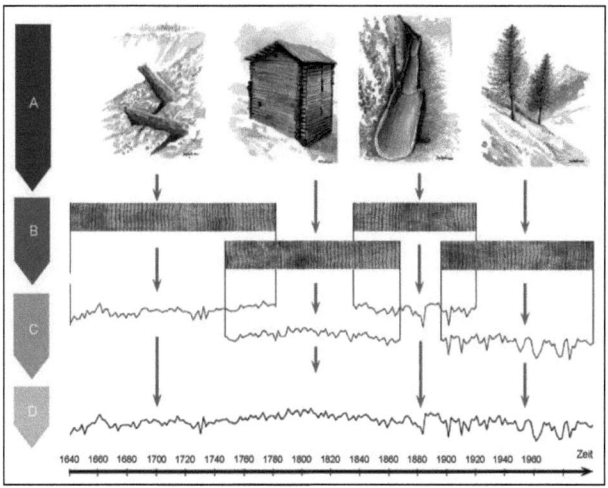

Fig. A4.1 Tree-ring research was originally used for dating purposes with crossdating methods. Growth rings from different trees were analyzed and their widths determined in order to obtain a continuous sequence that allows dating of wood samples of unknown age (Illustration: ETH 2009).

is called "cross-dating" and can be used to create long continuous tree-ring series (Schweingruber 1983). Resulting sequences then allow dating of wood of unknown age with yearly precision (Fig. A4.1).

Before the 1960s, tree-ring research was exclusively used for dating purposes in dendrochronology or dendroarcheology. Dendrogeomorphology evolved from the pure dating of wood to the much broader field of dendroecology, including all areas of science that deduce environmental data originating from tree-ring sequences (Schweingruber 1996). Dendrogeomorphology nowadays represents one subfield of dendroecology, methods were firstly described by Alestalo (1971) and further developed by Shroder (1980), Braam et al. (1987a,b), Butler (1987), and Shroder and Butler (1987). As presented, numerous internal and external factors continuously influence tree growth and were registered within the tree-rings. Likewise, many geomorphic processes have the potential to affect tree growth, such as rockfall, debris flows, landslides, avalanches, or erosion. The analysis of geomorphic processes through the study of growth anomalies in tree-ring series is called dendrogeomorphology (Alestalo 1971). A complete overview of current research in dendrogeomorphology can be found in Stoffel and Bollschweiler (2008, 2009).

4.2 TREES AND ROCKFALL

Tree-ring analyses of geomorphic processes are based on the concept of "process–event–response" presented by Shroder (1978). The disturbing agent, rockfall in the present case, represents the *process*. After the *event* (impact), the damaged tree will react upon disturbance with a characteristic growth *response*. In the next sections, the different events caused by rockfall will be presented as well as the responses illustrated.

4.2.1 WOUNDING (INJURIES)

If the impact energy of a falling rock is high enough to penetrate the bark, damage of the beneath vascular cambium (i.e. tissue that

Fig. A4.2 An injured stem of a Larix decidua tree.

Fig. A4.3 Cross-section of a wound, overgrowth starts at the lateral sides of the injury.

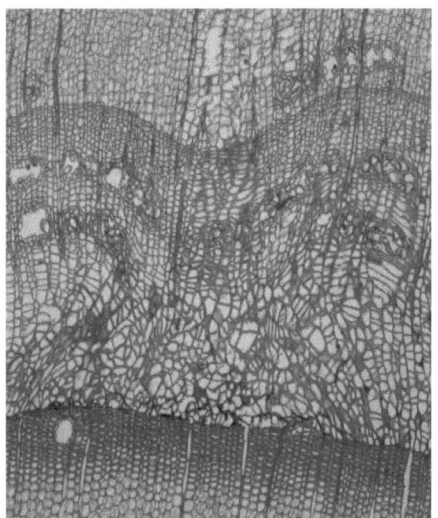

Fig. A4.4 Macroscopic view of callus tissue (Bollschweiler, 2007).

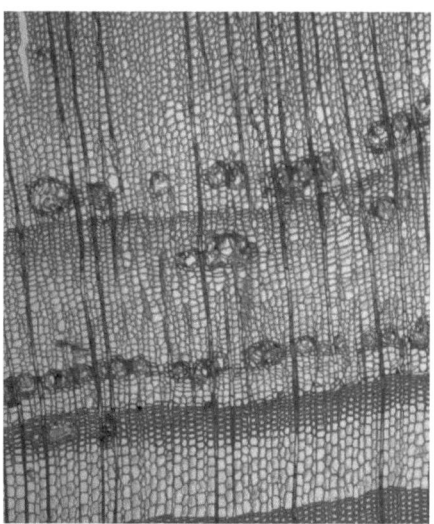

Fig. A4.5 Macroscopic view of TRD (Bollschweiler, 2007).

divides off any cells, "growth zone") occurs. At this position, cell-dividing is no longer possible, resulting in a section with no growth at all. In order to minimize the risk of rot and insect attack after wounding, the injured tree will compartmentalize the wood and almost immediately start with callus tissue formation (chaotic cell growth) at the edges of the injury (Shigo 1984, Sachs 1991, Larson 1994). To close the gaping wound, cambium cells start to overgrow continuously from the lateral edges of the wound (Stoffel 2005a, b, Stoffel and Perret 2006). Certain conifer species (*Larix decidua*, *Picea abies*, or *Abies alba*) additionally form tangential rows of traumatic resin ducts (TRD) are on both sides of the wound (Fahn et al. 1979, Nagy et al. 2000, Bollschweiler et al. 2008a). TRDs are differentiated immediately after the impact, their position can be used to date injuries even with intra-seasonal precision (Stoffel et al. 2005a, Stoffel 2008, Stoffel and Hitz 2008). Figure A4.2 illustrates the appearance of an injury in a living tree, Figure A4.3 shows a wound on a cross section. Figure A4.4 and Figure A4.5 picture microscopic representation of callus tissue and TRD.

4.2.2 BREAK-OFF OF BRANCHES AND STEM BREAKAGE

The total energy of a travelling boulder is determined by ist mass and ist velocity (translational and rotational). The impact angle then decides on the amount of energy released to the tree. If the impacting energy is too high for being absorbed by the tree, break-off of branches or even breakage of stems can occur. In general, breakage is more common in large trees where suppleness is lost. Breaking of branches does in general not severely affect tree growth, however the loss of big branches can potentially lead to a general growth decrease (Fig. A4.6). The consequences of stems breakage mainly depend on the height of collapse. Breakage close to ground level is in general a lethal injury, breaking of the top due to propagating sinusoidal shockwaves ("hula-hoop" effect, Dorren et al. 2006a) leads to distinct growth suppression in the years following the impact (Fig. A4.7, Fig. A4.8). In order to recover, one or several top branches try to replace the broken crown by growing in a vertical direction. This procedure results in a "candelabra" tree morphology (Butler and Malanson 1985, Shroder and Butler 1987, Fig. A4.9). Finally, the shock of the impact can induce the formation of TRD.

4.2.3 INCLINATION OF STEM

The released energy of an impacting rock is very directional and can lead to an inclination of the stem (Fig. A4.10). In order to regain its vertical position and stability, conifers start with the formation of reaction wood on the downslope side of the stem (Clauge and Souther 1982, Giardino et al. 1984, Braam et al. 1987a, b, Duncker and Spiecker 2008).

Fig. A4.7 Broken stem of a Picea abies tree after collision.

Fig. A4.8 Growth suppression in a tree-ring series after decapitation.

Fig. A4.6 Broken branch due to rockfall impact.

Fig. A4.9 Example of a "candelabra" tree. After breakage of its top, lateral branches took the lead and form a new crown.

Cell walls of tracheids become much more important (Timell 1986, Shroder 1980), leading to a darker color of reaction wood (Fig. A4.11). On tree ring series, eccentric growth can be observed after a tilting event. On the downslope side, tree rings become considerably wider than on the upslope side (Fig. A4.12). In some cases, additional growth suppression can be found on the upslope side, resulting in an increased difference of tree-ring width between the sides. Again, the shock of the impact can cause additional TRD formation.

Fig. A4.10 *Two trees were tilted by a falling rock, the fragment was stopped and remained in position.*

4.2.4 ELIMINATION OF NEIGHBORING TREES

Rockfall events of multiple fragments or of very large boulders can have devastating consequences, by elimination of entire forest stands. Remaining individuals growing next to cleared surfaces can benefit from better growth conditions (more light, nutrients or water) with less competition (Fig. A4.13). Ameliorated growth conditions result in growth release after clearance that often starts with a certain delay (Fig. A4.14).

Fig. A4.11 *Sudden appearance of darker reaction wood after tilting.*

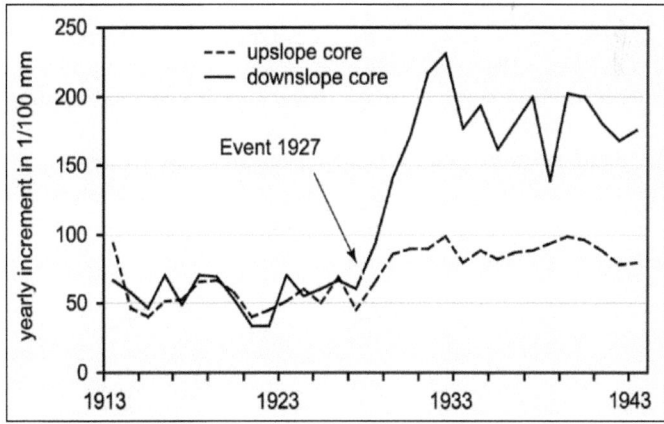

Fig. A4.12 *Measured tree-rings series of upslope (dashed line) and downslope sides reveal major differences of ring widths after inclination (Bollschweiler, 2007).*

4.3 METHODS IN DENDROGEOMORPHOLOLOGY

For all dendrogeomorphologic studies that aim at reconstructing previous activity of a process, it is crucial to start with a detailed overview of processes present at the study site. It has to be ensured, that the process of interest remains the only geomorphic process that influences trees. If several processes coexist (e.g., snow avalanches, rockfall, debris flows), it is not possible to assign observed growth reactions to the causing process.

Fig. A4.13 A large rockfall event in Zen Eisten (Valais, Switzerland) cleared major part of the forest stand, remaining trees benefit from better growth conditions.

4.3.1 LOCALIZATION

Dendrogeomorphic studies in rockfall research are mostly conducted either in the transit or the deposition zone of sites with rather small boulder sizes (big boulders would destroy the trees, rendering dendrogeomorphic studies impossible). The scree slopes of the transit zones as well as in the deposition zones present in general a homogenous surface. Therefore, no geomorphic mapping is necessary. However, sampled trees must be accurately located and positions transferred on a topographic map. In steep and forested valleys, GPS devices do often not give precise position. In order to determine the correct positions of sampled trees, (aerial) photographs or topographic sheets can be used. If the quality of available photographs is high, it is possible to localize single trees or particular objects, such as big boulders or anthropogenic constructions (Fig. A4.15, Fig. A4.16, Fig. A4.17, and Fig. A4.18). If these objects are identified in the field, they can serve as fixed points. Then positions can be mapped using tape measure, compass and inclinometer.

4.3.2 SAMPLING METHODS AND STRATEGIES

Trees showing obvious signs of disturbance induced by rockfall (scars, tilted stem, broken stems or tree tops) are selected for sampling. In order to study a larger area,

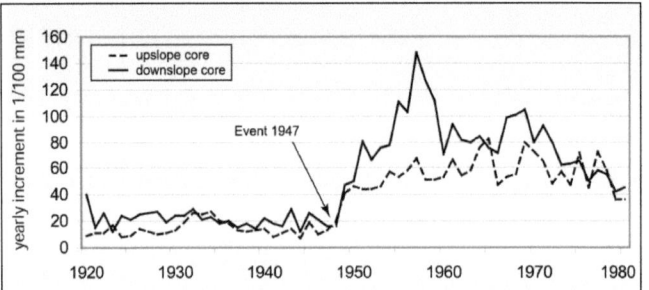

Fig. A4.14 Tree ring-series of a tree following clearance of neighboring individuals showing a sudden growth release of both sides (Bollschweiler, 2007).

4. Dendrogeomorphology - 43

Fig. A4.15 *Photograph of one of the study sites of present thesis. Picture taken from the opposite slope of the valley.*

Fig. A4.16 *Enlargement of the top section of the same study site, single trees or big boulders can be identified.*

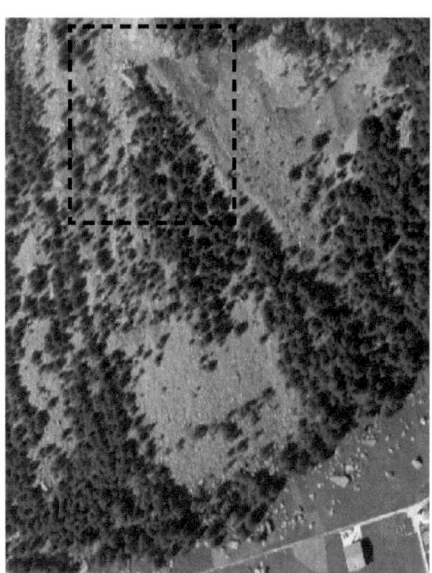

Fig. A4.17 *Identical location on an aerial photograph (Swisstopo 2009, [SwissImage 2005, © DV023268]).*

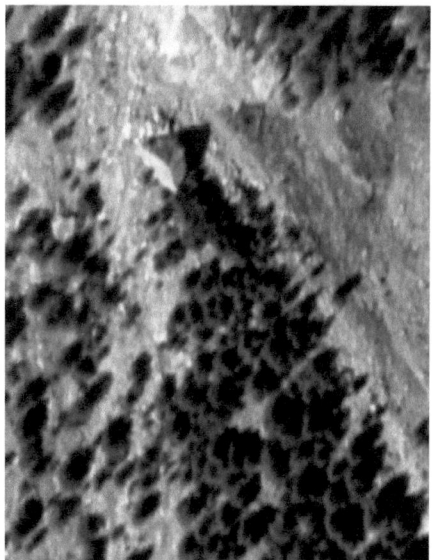

Fig. A4.18 *Corresponding extract of equal section (Swisstopo 2009, [SwissImage 2005, © DV023268]).*

trees can be sampled along horizontal or vertical transects with an equal distance to each other. The topographic position of the individual is determined, as well as its diameter at breast height, height, and information on neighboring trees. In a next step, description and sketches of all visible disturbances in the tree morphology are noted (scars, tilted stem axis, decapitation). Finally, pictures of the tree and each injury are taken in order to facilitate interpretation of laboratory results.

Trees then are felled with a handsaw or a chainsaw, depending on the diameter and number of samples (Fig. A4.19). Cross-sections at the heights of interest are taken and labelled. It has to be taken into consideration that extensive sampling results in a considerable amount of wood that has to be transported, prepared and analyzed. Depending on the aims of the study, tree characteristics, terrain, or permissions, it is not always possible or necessary to fell trees. In these cases, sampling can be performed with increment borers (Fig. A4.20). Obtained cores have a diameter of 5 mm. As core samples show only a fraction of the tree, sample position must be chosen with prudence (Fig. A4.21). More detailed information on used methods can be found in the methods part of Chapter B and C.

However, extraction of several cores per tree and injury is often necessary. In case of wound analysis, the sample is taken as close to the injury as possible where the vascular cambium has not been destroyed. Determination of best sampling position in practice can be difficult. If sampling an old injury, the wound edge is already overgrown from the sides and not visible anymore. Figure A4.22 shows resulting cores of different sampling positions with possible "loss" of several tree-rings. A more detailed description can be found in Stoffel and Bollschweiler (2008).

Fig. A4.19 Sampling of a tree with the chainsaw.

Fig. A4.20 Sample extracting by the use of an increment borer.

Fig. A4.21 Representation of tree sampled with an increment borer before a cross-section was taken.

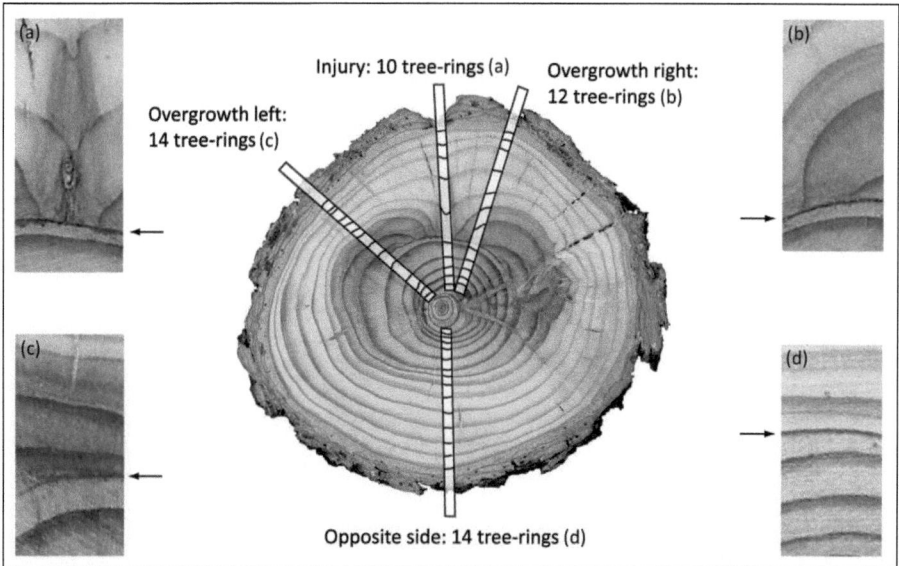

Fig. A4.22 *Illustration of wound sampling with an increment borer at different positions with resulting cores. As can be seen, sampling too close or too far away from the injury leads to missing years or loss of growth reaction. The year of wounding is indicated by the black arrows.*

Fig. A4.23. *Example of an insect outbreak. Caterpillar of the Larch bud moth (Zeiraphera Diniana Gn., Photo courtesy by CFL 2003).*

Fig. A4.24 *Destroyed young buds of Larix decidua trees after infestation (Photo courtesy by CFL 2003).*

Fig. A4.25 *Adult moths after pupation (Photo courtesy by Werner Baltensweiler).*

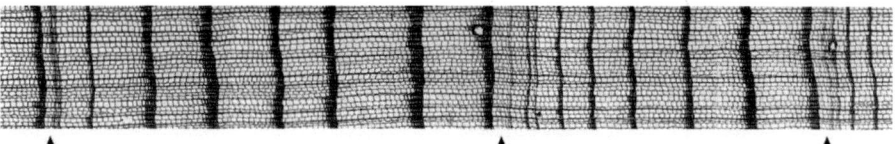

Fig. A4.26 *Tree-ring series reflecting regular outbreaks of the Larch bud moth (black arrows, Schweingruber 2001).*

4.3.3 SAMPLING PREPARATION AND ANALYSIS

In order to make the tree rings and cell structures visible, all samples need to be air dried and polished with 400 grit sand paper. For sanding, core samples are glued on a woody support. Thereby, wood fibres need to be in vertical position. Further explications of sample preparation can be found in Iseli and Schweingruber (1989).

If the samples are prepared, tree rings are counted, starting from the outermost ring. Thereafter, tree-rings widths are measured, using a LINTAB positioning table, coupled to a Leica stereomicroscope. For the present study, the TSAP (Time Series Analyses and Presentation, Rinntech 2009) software was used for measuring tree widths with an accuracy of 1/100 mm.

In trees that are not disturbed by geomorphic processes, growth suppression or release are driven by climatic conditions or by insect outbreaks (Fig. A4.23, Fig. A4.24, Fig. A4.25, and Fig. A4.26). Trees of the same species growing at the same site are influenced by the same factors and are therefore supposed to show the same growth pattern. The general growth pattern of a species at a specific site can be summarized in a so called "reference chronology" (Cook and Kairiukstis 1990, Schweingruber 1996).

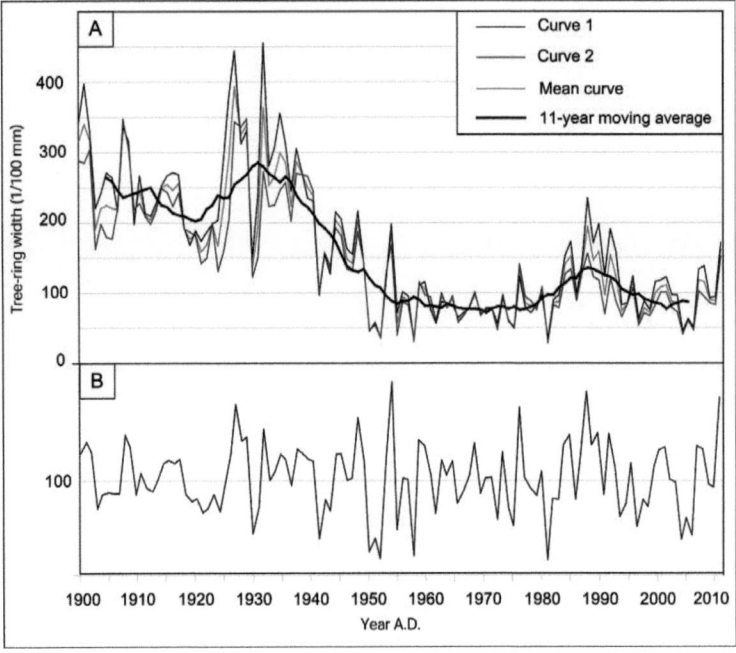

Fig. A4.27 *(a) Illustration of tree ring series of two sides of a reference tree (black and blue curve) with its mean (red curve). The black bold curve represents the 11-year moving average serving for long-term trend removal. (b) Resulting curve after indexation, the mean of several indexed curves would lead to a reference chronology (Illustration: Bollschweiler 2007, adapted).*

To obtain a reference chronology, undisturbed trees growing at the same site are sampled, and ring widths measured. Then, mean curves are constructed for each tree and long-term trends such as aging are removed by standardizing the resulting curve with an 11-year moving average (Cook and Kairiukstis 1990). Trees showing abnormal growth pattern are then removed from further processing. Good quality series are indexed in the next step before an overall mean curve is constructed, resulting in the reference chronology. Figure A4.27 illustrates tree ring series of two sides of a reference tree as well as following the standardization and indexation procedure.

4.3.4 DATING ROCKFALL EVENTS

To identify all growth reactions resulting from rockfall activity, tree-ring series of disturbed trees can be compared to the corresponding growth sequence of the reference chronology. This method allows to determine if observed changes in growth were caused by general "natural" conditions and if tree rings are entirely missing (e.g. insect outbreak, drought). If a conspicuous growth pattern can not be found within the reference chronology, more specific events, such as geomorphic processes, must have influenced tree growth. Growth anomalies induced by rockfall impacts include formation of callus tissue, differentiation of TRD, growth suppression, reaction wood, and growth release. The years of growth anomalies are noted and, in case of cross-sections, the intra-annual position of reactions within the season.

As core sampling during this study was performed exclusively next to visible wounds, it can be assured that present reactions were the result of disturbances caused by falling rocks (other scar-inducing processes such as debris flows or snow avalanches should not occur on selected study sites). Growth anomalies had to be clearly present in order to be noted. In case of TRD, the limitations defined by Stoffel et al. (2005b) were additionally applied. To be considered, TRD had to be: (i) traumatic, (ii) extremely compact, and (iii) form continuous rows.

4.4 DENDROGEOMORPHIC RESEARCH – A SHORT OVERVIEW

Many geomorphic processes have been studied using dendrogeomorphic methods. In the following, a short overview is given, based on an overview paper presented by Stoffel and Bollschweiler (2008). The first dendrogeomorphic studies investigated *erosion* rates using tree roots (La Marche 1961, 1968, Carrara and Carroll 1979). Erosion is still subject of several recent dendrogeomorphic studied published by Bodoque et al. (2005), McAuliffe et al. (2006), Scuderi et al. (2008), Rubiales et al. (2008), or Hitz et al. (2008). *Debris flows* are one of the most intensely investigated mass-movement processes in dendrogeomorphology. Early studies were conducted in the 1980s by Hupp (1984), Hupp et al. (1987) or Strunk (1989, 1991, 1997). More recently, several studies were published in Switzerland, namely by Baumann and Kaiser (1999), Stoffel et al. (2005c, 2008a, b), Bollschweiler and Stoffel (2007) Bollschweiler et al. (2007, 2008a, b). The tree ring based analysis of *landslides* was studied firstly by Bégin and Filion (1985, 1988), Braam (1987a, b) and Filion et al. (1991). More recent studies that investigated landslides were aiming for the identification of the main debris flow triggers (Corominas and Moya 1999, Fantucci and McCord 1995, Fantucci and Sorriso-Valvo 1999). Several other processes have been subject to numerous tree-ring based studies, such as *snow avalanches* (Butler 1979, 1985, Boucher et al. 2003, Dubé et al. 2004, Muntán et al. 2004, 2008, Stoffel et al. 2006b, Mundo et al. 2007, Butler and Sawyer 2008, Casteller et al. 2008) or *earthquakes* (Jacoby et al. 1988, 1992, Jacoby 1997).

As seen, there exist numerous fields in geomorphic research, where tree-rings were successfully used in order to investigate characteristics of various processes. However, dendrogeomorphic studies in *rockfall* research remain very scarce. Initially, tree-rings were used to identify and date large rock avalanches (Moore and Matthews 1978, Butler et al. 1986). Lafortune et al. (1997) focused on small-scale rockfall events, but aiming for the reconstruction of sedimentation rates and forest edge dynamics. Stoffel et al. (2005a, b) were the first who revealed the potential of dendrogeomorphic methods in rockfall research. They reconstructed four entire centuries of rockfall activity as well as the intra seasonal distribution of events. Perret et al. (2006) likewise reconstructed rockfall activity for the last 120 years, focusing on possible meteorological triggers (precipitation, seasonal temperatures). Finally, Stoffel et al. (2006a) used tree-ring provided data in order to verify outcomes of a three-dimensional rockfall simulation model. A comprehensive overview of investigations dealing with tree-ring sequences and rockfall activity is presented in Stoffel (2006).

♦♦♦♦♦

5 STUDY SITES

Fieldwork for the present thesis was conducted on three different sites in the Alps (Fig. A5.1). Two of the study sites are located in the southern Valais Alps, namely at Saas Balen and at Täsch. The third site investigated is located near Vaujany (Isère, French Alps). Studies of Chapter B1, C1 and C2 were exclusively conducted with material from Saas Balen. The study sites of Chapter B2 were located in Saas Balen, Täsch, and Vaujany.

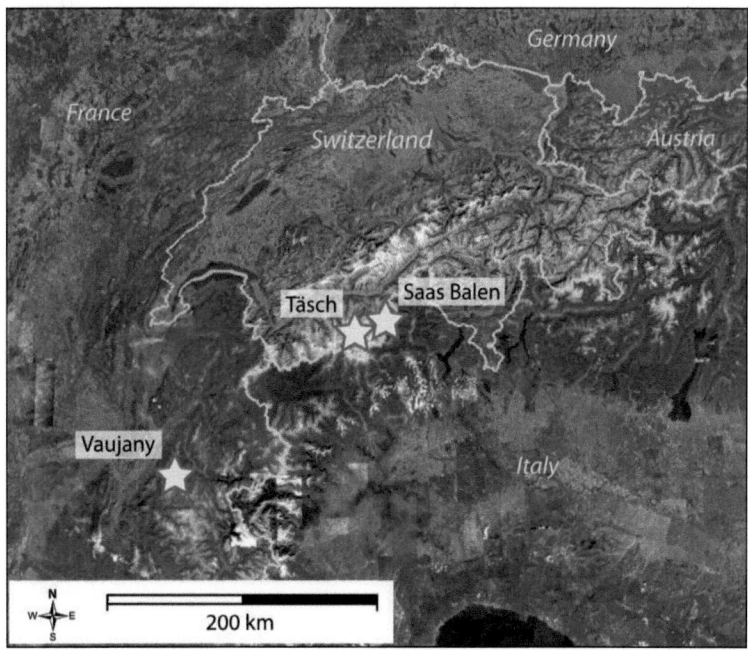

Fig. A5.1 Study sites of the present thesis are located in the southern Valais and the French Alps (red stars, Google Earth 2009, adapted).

5.1 Saas Balen

The trees used for the analyses in Chapter B1, C1, and C2 originated from a forest stand called "Schilt" near Saas Balen (Valais, Switzerland, 46°09' N., 7°55' E.). The *Picea abies* individuals of Chapter B2 were sampled in Saas Balen as well (Fig. A5.2 - Fig. A5.4).

The forest stand investigated is located between 1390 and 1610 m asl on the ENE-facing slope below the Lammenhorn (3189 m asl). Bedrock in the zone consists of micaceous schists belonging to Penninic crystalline layers, dipping SSE with an angle of about 20° (Bearth 1973, 1980). Rockfall originates from the disintegrated and glacially oversteepened cliffs at 1750–1900 m asl (Fig. A5.2). The mean slope angle is 36° with only a small variation between the top and the bottom of the study site (Fig A5.3). The volume of falling rocks does not normally exceed 1 m^3 on the study site. However, a few blocks with volumes of up to 50 m^3 are deposited in the valley bottom, bearing witness to major events in the past. Archival data and local toponomy indicate the occurrence of rockfall in the region since at least the early eighteenth century when rockfalls descended from the neighboring "Steinschlagwald" (= rockfall forest) and destroyed the old parish church (Ruppen et al. 1979). In contrast, other mass-movement processes, such as debris flows or snow avalanches, have never been witnessed on the slope.

Quaternary talus and morainic deposits cover the transition and deposition zones. A centimetric soil layer covers rockfall depos-

Fig. A5.3 Investigated forest with a mean slope angle of 36°.

Fig. A5.2 Photograph of the study site in Saas Balen, Valais, Switzerland (white form).

Fig. A5.4 View from inside the forest stand.

its within the study site, whereas the surfaces outside the study area remain almost free of vegetation and are covered with bare rocks and boulders. The stand at Schilt mainly consists of *Larix decidua* trees (90%), of some *Picea abies* (10%), and single *Pinus cembra* trees (Fig. A5.4). Three separate forest units exist; the largest located in the upper half of the site is covered with a stand that becomes gradually denser towards its upper limits. The two other units, only considered in Chapter C3, each consist of small forested bands oriented perpendicular to the fall line and are located at the bottom of the investigated area. No anthropogenic influence is visible in the studied or neighboring forest stand.

5.2 Täsch

The *Larix decidua* trees of Chapter B3 were collected in Täsch (Valais, Switzerland, 46°04' N., 7°47' E.), at an altitude of 1800 – 1850 m asl. The trees sampled were located at the west-facing Täschgufer slope, descending from the Leiterspitzen summit (3214 m asl, Fig. A5.5, Fig. A5.6) in the Siviez–Mischabel nappe. Layers of the bedrock generally strike SSW and dip WNW with angles of 40–80° (Lauber 1995).

Fig. A5.6 Aerial photograph of the study site (white form), located above a rockfall protection dam (Swisstopo 2009, [SwissImage 2005, © DV023268]).

Fig. A5.5 General view of the study site at the Täschgufer slope in Valais, Switzerland (white form).

Fig. A5.7 View from inside the forest stand, note the smashed trees.

Rockfall frequently occurs on the slope, originating from the heavily disintegrated gneissic rockwalls. There exist two main rockfall source areas on the slope. The first source area is located between 2300 and 2600 m asl, the second rockfall generating area can be found in a zone of highly fractured bedrock with many joints above 2700 m asl. The volume of single rockfall fragments normally does not exceed 2 m³. The mean slope angle is 35°, with a maximum value of 48° in the upper part and gradually decreasing angles to only 20° near the valley floor.

The forest at the study site consists exclusively of *Larix decidua* trees with large gaps between the individuals (Fig. A5.7). In the past, rockfall regularly caused damage to roads and hiking trails. In order to protect human infrastructure from damage, large rockfall protection dams were built between 1988 and the late 1990s.

5.3 VAUJANY

The study site of the *Abies alba* trees used in Chapter B2 is located in Vaujany (Isère, France, 45°12' N., 6°03' E) with an altitude of 1300 – 1400 m asl. The boulder sizes of the NW facing rockfall slope with a mean slope angle of 38° do not exceed 1m³. Bedrock at the study site consists of granite belonging to the Sept-Laux massif. The granitic scree slope is covered by a thin soil layer of variable thickness. The forest consists mainly of four different species: *Abies alba* (50%), *Picea abies* (25%), *Fagus sylvatica* L. (European Beech, 17%), and *Acer pseudoplatanus* L. (Sycamore Maple, 4%, Bourqui 2009)

In addition to natural rockfall, the site was used for real-size rockfall experiments in order to analyze rockfall-forest interactions ("RockFor" project, Dorren et al. 2006b). Trees investigated in the present study were impacted by natural and artificially triggered rockfall. The general aspects of the study site in Vaujany can be seen in Figure A5.8 and A5.9.

Fig. A5.8 General view of the study site in Vaujany, the position of sampled trees is indicated by the white form (Photo courtesy by Karin Bourqui).

Fig. A5.9 View from inside the forest (Photo courtesy by Markus Stoffel).

✦✦✦✦✦

6 BIBLIOGRAPY OF CHAPTER A

Abele, G., 1971. Bergstürze in den Alpen. Wissenschaftliche Alpenvereinshefte 25. München. (In German).

Agliardi, F., Crosta, G.B., 2003. High resolution three-dimensional numerical modelling of rockfalls. International Journal of Rock Mechanics and Mining Sciences 40, 455–471.

Agliardi, F., Crosta, G., Zanchi, A., 2001. Structural constraints on deep-seated slope deformation kinematics. Engineering Geology 59, 83–102.

Alestalo, J., 1971. Dendrochronological interpretation of geomorphic processes, Fennia, 105, 1–139.

Allaby, A., Allaby, M., 1990. Oxford concise dictionary of earth sciences. New York: Oxford University Press, p. 410.

André, M.F., 1986. Dating slope deposits and estimating rates of rockwall retreat in northwest Spitsbergen by lichenometry. Geografiska Annaler 68A, 65–75.

ASTRA, 2009 (Bundesamt für Strassen). http://www.astra.admin.ch/themen/nationalstrassen/01294/index.html?lang=de (as seen on 26 March 2009).

Ayala-Carcedo, F. J., Cubillo-Nielson, S., Alvarez, A., Dominguez, M. J., Lain, J., Lain, R., Ortiz, G., 2003. Large scale rockfall reach susceptibility maps in La Cabrera Sierra (Madrid) performed with GIS and dynamic analysis at 1:5,000. Natural Hazards 30, 325–340.

Azzoni, A., Rossi, P.P., 1991. *In situ* observation of rockfall analysis parameters. Landslides, Bell (ed.). Balkema, Rotterdam, pp. 307-314.

Azzoni, A., Barbera, G. L., Zanizetti, A., 1995. Analysis and prediction of rockfalls using a mathematical model. International Journal of Rock Mechanics and Mining Science 32, 709–724.

Badger, T.C., Lowell, S.M., 1992. Rockfall control in Washington State. In: Rockfall Prediction and Control and Landslide Case Histories. Transportation Research Record 1343, National Academy of Sciences, Washington, D.C., pp 14-19.

Baillifard, F., Jaboyedoff, M., Sartori, M., 2003. Rockfall hazard mapping along a mountainous road in Switzerland using a GIS-based parameter rating approach. Natural Hazards and Earth System Sciences 3, 431–438.

Bajgier-Kowalska, M., 2002. The application of lichenometry in the dating of landslide-rockfall slopes in the Beskid Zywiecki Mountains (Flysh Carpathians). Czasopismo Geograficzne 73, 215–230.

Bassato, G., Cocco, S., Silvano, S., 1985. Programma di simulazione per lo scoscendimento di blocchi rocciosi. Dendronatura 6 (2), 34–36. (In Italian).

Baumann, F., Kaiser, K.F., 1999. The Multetta debris fan, eastern Swiss Alps: a 500-year debris flow chronology. Arctic and Alpine Research 31, 128–134, 1999.

Bearth, P., 1973. Geologischer Atlas der Schweiz 1:25000, Simplon (Atlasblatt 61). Schweizerische Geologische Kommission. (In German).

Bearth, P., 1980. Geologischer Atlas der Schweiz 1:25000, Niklaus (Atlasblatt 71). Schweizerische Geologische Kommission. (In German).

Bebi, P., Kienast, F., Schönenberger, W., 2001. Assessing structures in mountain forests as a basis for investigating the forests dynamics and protective function. Forest Ecology and Management 145, 3–14.

Becker, A., Davenport, C., 2005. Rockfalls triggered by the AD 1356 Basle Earthquake. Terra Nova 15, 258–264.

Bégin, C., Filion, L., 1985. Analyse dendrochronologique d'un glissement de terrain dans la region du Lac de l'Eau Claire (Québec nordique). Canadian Journal of Earth Sciences 22, 175–182. (In French)

Bégin, C., Filion, L., 1988. Age of landslides along Grande Rivière de la Baleine estuary, eastern coast of Hudson Bay, Quebec (Canada). Boreas 17, 289–299.

Berger, F., Quetel, C., Dorren, L.K.A., 2002. Forest: A natural protection mean against rockfall, but with which efficiency? The objectives and methodology of the ROCKFOR project. In: Proceedings International Congress Interpraevent 2002 in the Pacific Rim, Matsumoto, Japan, pp. 815–826.

Beylich, A.A., Sandberg, O., 2005. Geomorphic effects of the extreme rainfall event of 20–21 July, 2004 in the Latnjavagge catchment, northern Swedish Lapland. Geografiska Annaler 87A, 409–419.

Bjerrum, L., Jorstad, F., 1966. Stability of rock slopes in Norway. Norwegian Geotechnical Institute 67, 59–78.

Blackwelder, E., 1942. The process of mountain sculpturing by rolling debris. Journal of Geomorpology 4, 324–328.

Bollschweiler, M., 2007. Spatial and temporal occurrence of past debris flows in the Valais Alps - results from tree-ring analysis. PhD thesis. Faculty of Science, University of Fribourg. GeoFocus 20, 1–182.

Bollschweiler, M., Stoffel, M., 2007. Debris flows on forested cones – reconstruction and comparison of frequencies in two catchments in Val Ferret, Switzerland. Natural Hazards and Earth System Science 7, 207–218.

Bollschweiler, M., Stoffel, M., Ehmisch, M., Monbaron, M., 2007. Reconstructing spatio-temporal patterns of debris-flow activity with dendrogeomorphological methods. Geomorphology, 87(4), 337–351.

Bollschweiler, M., Stoffel, M., Schneuwly, D.M., Bourqui, K., 2008a. Traumatic resin ducts in *Larix decidua* stems impacted by debris flows. Tree Physiology 28, 255–263.

Bollschweiler, M., Stoffel, M., Schneuwly, D.M., 2008b. Dynamics in debris-flow activity on a forested cone – a case study using different dendroecological approaches. Catena 72, 67–78.

Bodoque, J.M., Díez-Herrero, A., Martín-Duquea, J.F., Rubiales, J.M., Godfrey, A., Pedraza, J., Carrasco, R.M., Sanz, M.A., 2005. Sheet erosion rates determined by using dendrogeomorphological analysis of exposed tree roots: Two examples from Central Spain. Catena 64, 81–102.

Boucher, D., Filion, L., Hétu, B., 2003. Reconstitution dendrochronologique et fréquence des grosses avalanches de neige dans un couloir subalpin du mont Hog's Back, en Gaspésie centrale (Québec), Géographie Physique et Quaternaire 57, 159–168. (In French).

Bourqui, K., 2009. Chutes de pierres et croissance des arbres: les réactions d'un arbre sur l'impact des chutes de pierres ? Master Thesis, Department of Geoscience, Univeristy Fribourg, Switzerland. (In French).

Bozzolo, D., Pamini, R., 1986a. Simulation of rock falls down a valley side. Acta Mechanica 63, 113–130.

Bozzolo, D., Pamini, R., 1986b. Modello matematico per lo studio della caduta dei massi. Laboratorio di Fisica Terrestre ICTS. Dipartimento Pubblica Educazione, Lugano-Trevano, 89pp. (In Italian).

Bozzolo, D., Pamini, R., Hutter, K., 1988. Rockfall analysis – a mathematical model and its test with field data. In: Proceedings of the 5th International Symposium on Landslides, Lausanne, pp. 555–560.

6. Bibliography of Chapter A - 55

Braam, R.R., Weiss, E.E.J., Burrough, A., 1987a. Spatial and temporal analysis of mass movement using dendrochronology. Catena 14, 573–584.

Braam, R.R., Weiss, E.E.J., Burrough, A., 1987b. Dendrogeomorphological analysis of mass movement: A technical note on the research method. Catena 14, 585–589.

Braathen, A., Blikra, L.H., Berg, S.S., Karlsen, F., 2004. Rock-slope failures in Norway; type, geometry, deformation mechanisms and stability. Norwegian Journal of Geology 84, 67–88.

Brang, P., 2001. Resistance and elasticity: promising concepts for the management of protection forests in the European Alps. Forest Ecology and Management 145, 107–117.

Brang, P., Schönenberger, W., Ott, E., 2000. Forests as Protection From Natural Hazards, The Forests Handbook, vol. 2. Blackwell Scientific, Oxford.

Brang, P., Schönenberger, W., Ott, E., Gardner, B., 2001. Forests as protection from natural hazards. In: EVANS, J. (ed.), The Forests Handbook Volume 2. Blackwell Science, Oxford, pp. 53–81.

Brang, P., Schönenberger, W., Frehner, M., Schwitter, R., Thormann, J.J., Wasser, B., 2006. Management of protection forests in the European Alps: an overview. Forest Snow and Landscape Research 80(1), 23–44.

Brauner, M., Weinmeister, W., Agner, P., Vospernik, S., Hoesle, B., 2005. Forest management decision support for evaluating forest protection effects against rockfall. Forest Ecology and Management 207, 75–85.

Broilli, L., 1974. Ein Felssturz in Grossversuch. Rock Mechanics, Supplementum 3, 69–78.

Budetta, P., 2004. Assessment of rockfall risk along roads. Natural Hazards and Earth System Sciences 4, 71–81.

Budetta, P., Santo, A., 1994 Morphostructural evolution and related kinematics of rockfalls in Campania (southern Italy): a case study. Engineering Geology 36, 197–210.

Bull, W. B., Brandon, M. T., 1998. Lichen dating of earthquake-generated regional rockfall events, Southern Alps, New Zealand. Geological Society of America Bulletin 110, 60–84.

Bunce, C.M., Cruden, D.M., Morgenstern, N.R., 1997. Assessment of the hazard of rock fall on a highway. Canadian Geotechnical Journal 34, 344–356.

Butler, D.R., 1979. Snow avalanche path terrain and vegetation, Glacier National Park, Montana. Arctic, Antarctic, and Alpine Research 11, 17–32.

Butler, D.R., 1985. Vegetational and geomorphic change on snow avalanche paths, Glacier National Park, Montana, U.S.A. Great Basin Naturalist 45(2), 313-317.

Butler, D.R., 1987. Teaching general principles and applications of dendrogeomorphology, Journal of Geological Education 35, 64–70.

Butler, D.R., 1990. The geography of rockfall hazards in Glacier National Park, Montana. Geographical Bulletin – Gamma Theta Upsilon 32(2), 81–88.

Butler, D.R., Malanson, G.P., 1985. A history of high-magnitude snow avalanches, southern Glacier National Park, Montana, USA. Mountain Research and Development 5, 175-182.

Butler, D.R., Oelfke, J.G., Oelfke, L.A., 1986. Historic rockfall avalanches, northeastern Glacier National Park, Montana, U.S.A. Mountain Res. Dev. 6, 261–271.

Butler, D.R., Sawyer C.F., 2008. Dendrogeomorphology and High Magnitude Snow Avalanches: A Review and Case Study. Natural Hazards and Earth System Sciences 8, 303-309.

BUWAL, BRP, BWW, 1997. Naturgefahren. Empfehlungen 1997. Berücksichtigung der Massenbewegungsgefahren bei raumwirksamen Tätigkeiten. Bundesamt für Umwelt, Wald und Landschaft, Bundesamt für Raumplanung und Bundesamt für Wasserwirtschaft, Bern. (In German).

Carrara, P.E., Carroll, T.R., 1979. The determination of erosion rates from exposed tree roots in the Piceance Basin, Colorado. Earth Surface Processes and Landforms 4, 307–317.

Case, W.F., 1988. Geological effects of the 14 and 18 August, 1988 Earthquake in Emery County, Utah. Utah Geological and Mineral Survey Notes 22, 8–14.

Casteller, A., Christen, M., Villalba, R., Martínez, H., Stöckli, V., Leiva, J., Bartelt, P., 2008. Validating numerical simulations of snow avalanches using dendrochronology: The Cerro Ventana event in Northern Patagonia, Argentina. Natural Hazards and Earth System Sciences 8, 433–443.

Chau, K.T., Wong, R.H.C., Lee, C.F., 1998. Rockfall problems in Hong Kong and some new experimental results for coefficient of restitution. International Journal of Rock Mechanics and Mining Science 35(4), 662–63.

Chau, K.T., Wong, R.H.C., Wu, J.J., 2002. Coefficient of restitution and rotational motions of rockfall impacts. International Journal of Rock Mechanics and Mining Sciences 39(1), 69–77.

Chau, K.T., Tang, Y.F., Wong, R.H.C., 2004. GIS based rockfall hazard map for Hong Kong. International Journal of Rock Mechanics and Mining Science 35(4), 662–663.

Chauvin, C., Renaud, J.P., Rupé, C., 1994. Stabilité et gestion des forêts de protection. Office Nationale des forêts, Bulletin Technique 27, 37–52. (In French).

Chen, H., Chen, R., Huang, T., 1994. An application of an analytical model to a slope subject to rockfall. Bulletin of the Association of Engineering Geologists 31(4), 447–458.

CFL (Centre de foresterie des Laurentides), 2003. http://www.cfl.scf.rncan.gc.ca/collections-cfl/ficheUinUsecte_e.asp?id=11213 (as seen on 10 May 2003).

Clague, J.J., Souther, J.G., 1982. The Dusty Creek landslide on Mount Caylay, British Columbia. Canadian Journal of Earth Sciences 19, 524-539.

Colin, K., Ballantyne, J., Stone O., 2004. The Beinn Alligin rock avalanche, NW Scotland: cosmogenic 10Be dating, interpretation and significance. The Holocene 14(3), 448–453.

Cook, E.R., Kairiukstis, L.A., 1990. Methods of dendrochronology – applications in the environmental sciences. Kluwer, London.

Corominas, J., Moya, J., 1999. Reconstructing recent landslide activity in relation to rainfall in the Llobregat River basin, Eastern Pyrenees, Spain. Geomorphology 30, 79–93.

Corominas, J., Ramon, C., José, M., Joan M.V., Joan Altimir, Jordi Amigó. 2005. Quantitative assessment of the residual risk in a rockfall protected area. Landslides 2, 343–357.

Couvreur, S., 1982. Les forêts de protection contre les risques naturels. Nancy: École Nationale du Génie Rural des Eaux et Forêts, p.89. (In French).

Crosta, G.B., Agliardi, F., 2003. A methodology for physically based rockfall hazard assessment. Natural Hazards and Earth System Sciences 3, 407–422.

Crosta, G.B., Agliardi, F., 2004. Parametric evaluation of 3D dispersion of rockfall trajectories Natural Hazards and Earth System Sciences 4, 583–598.

Crosta, G.B., Agliardi, F., Frattini, P., Imposimato, S., 2004. A three dimensional hybrid numerical model for rockfall simulation. Geophysical Research Abstracts 6, 04502.

Davies, M.C.R., Hamza, O., Harris, C., 2001. The effect of rise in mean annual temperatures on the stability of rock slopes containing ice filled discontinuities. Permafrost and Periglacial Processes 12, 137–144.

Decaulne, A., Sæmundsson, Þ., 2006: Meteorological conditions during slush-flow release and their geomorphological impact in northwestern Iceland: a case study from the Bíldudalur valley. Geografiska Annaler 88A(3), 187–197.

Descoeudres, F., 1990: L'éboulement des Crétaux: Aspects géotechniques et calcul dynamique des chutes de blocs. Mitteilungen der Schweizerischen Gesellschaft für Boden und Felsmechanik. Frühjahrstagung des 21.6.1990 in Sion. (In French).

Descoeudres, F., Zimmermann, T., 1987. Threedimensional dynamic calculation of rockfalls. In: Hegert, G., Vongpaisal, S., (eds.). Proceedings of the Sixth International Congress on Rock Mechanics, Montreal. Balkema, Rotterdam, pp. 337–342.

Donzé, F., Magnier, S.A., Montani, S., Descoeudres, F., 1999. Numerical study of rock block impacts on soil-covered sheds by a discrete element method. 5ème Conférence Nationale en Génie Parasismique, AFPS'99, Cachan. (In French)

Dorren, L.K.A., 2003. A review of rockfall mechanics and modelling approaches. Progress in Physical Geography 27(1), 69–87.

Dorren, L.K.A., 2008. Rockfall and protection forests – models, experiments and reality. Habilitation BOKU Vienna: 312 pp.

Dorren, L.K.A., Seijmonsbergen, A.C., 2003. Comparison of three GIS-based models for predicting rockfall runout zones at a regional scale. Geomorphology 56, 49–64.

Dorren, L.K.A., Berger, F., 2006. Stem breakage of trees and energy dissipation at rockfall impacts. Tree Physiology 26, 63–71.

Dorren, L.K.A., Maier, B., Putters, U.S., Seijmonsbergen, A.C., 2004a. Combining field and modelling techniques to assess rockfall dynamics on a protection forest hillslope in the European Alps. Geomorphology 57, 151–167.

Dorren, L.K.A., Berger, F., Imeson, A.C., Maier, B., Reya, F., 2004b. Integrity, stability and management of protection forests in the European Alps. Forest Ecology and Management 195, 165–176.

Dorren, L.K.A., Berger, F., Le Hir, C., Mermin, E., Tardif, P., 2005. Mechanisms, effects and management implications of rockfall in forests. Forest Ecology and Management 215, 183–195.

Dorren, L.K.A., Berger, F., Putters, U.S., 2006a. Real size experiments and 3D simulation of rockfall on forest slopes. Natural Hazards and Earth System Sciences 6, 145–153.

Dorren, L., Berger, F., Mermin, E., Tardif, P., 2006b. Results of Real Size Rockfall Experiments on Forested and Non-Forested Slopes. Disaster Mitigation of Debris Flows, Slope Failures and Landslides, pp. 223–228.

Dorren, L., Berger, F., Jonsson, M., Krautblatter, M., Mölk, M., Stoffel, M., Wehrli, A., 2007. State of the art in rockfall – forest interactions. Schweizerische Zeitschrift für Forstwesen 158(6), 128–141.

Douglas, G.R., 1980. Magnitude frequency study of rockfall in Co. Antrim, N. Ireland. Earth Surface Processes and Landforms 5(2), 123–129.

Dramis, F., Govi, M., Guglielmin, M., Mortara, G., 1995. Mountain permafrost and slope instability in the Italian Alps. The Val Pola Landslide. Permafrost and Periglacial Processes 6, 73–81.

Dubé, S., Filion, L., Hétu, B., 2004. Tree-ring reconstruction of high-magnitude snow avalanches in the Northern Gaspé Peninsula, Québec, Canada. Arctic, Antarctic, and Alpine Research 36, 555–564.

Duncker, P., Spiecker, H., 2008. Cross-sectional compression wood distribution and its relation to eccentric tangential growth in *Picea abies* [L.] Karst. Dendrochronologia 26, 195–202.

Dussauge-Peisser, C., Helmsdetter, A., Grasso, J.R., Hantz, D., Desvarreux, P., Jeannin, M., Giraud, A., 2002. Probabilistic approach to rock fall hazard assessment: potential of historical data analysis. Natural Hazards and Earth System Sciences 2, 15–26.

Easterbrook D.J., 1993. Surface Processes and Landforms. Department of Geology Western Washington University. Macmillan Publishing Company, p. 510.

Erismann, T.H., Abele, G., 2001. Dynamics of rockslides and rockfalls. Berlin, Springer, p. 316.

ETH 2009: http://www.fe.ethz.ch/lab/CrossdatUinUg (as seen on 12 March 2009).

Evans, S.G., Hungr, O., 1993. The assessment of rockfall hazards at the base of talus slopes. Canadian Geotechnical Journal 30, 620–636.

Evans, S.G., 1997. Fatal landslides and landslide risk in Canada. In: Proceedings of the International Workshop on Landslide Risk Assessment, Honolulu, USA. Balkema, Rotterdam, pp. 620–636.

Fahey, B.D., Lefebvre, T.H., 1988. The freeze–thaw weathering regime at a section of the Niagara Escarpment on Bruce Peninsula, Canada. Earth Surface Processes and Landforms 13, 293–304.

Fahn, A., Werker, E., Ben-Tzur, P., 1979. Seasonal

effects of wounding and the growth substances on development of traumatic resin ducts in *Cedrus libani*. New Phytologist 82, 537–544.

Falcetta, J.L., 1985. Un nouveau mod"ele de calcul de trajectoires de blocs rocheux. Revue Francaise de Geotechnique 30, 11–17. (in French).

Fantucci, R., McCord, A., 1995. Reconstruction of landslide dynamic with dendrochronological methods. Dendrochronologia 13, 43–58.

Fantucci, R., Sorriso-Valvo, M., 1999. Dendrogeomorphological analysis of a slope near Lago, Calabria (Italy). Geomorphology 30, 165–174.

Filion, L., Quinty, F., Bégin, Y., 1991. A chronology of landslide activity in the valley of Rivière du Gouffre, Charlevoix, Quebec. Canadian Journal of Earth Sciences 28, 250–256.

Foetzki, A., Jonsson, M., Kalberer, M., Simon, H., Mayer, A.C., Lundström, T., Stöckli, V., Ammann, W.J., 2004. Die mechanische Stabilität von Bäumen: das Projekt Baumstabilität des FB Naturgefahren. Schutzwald und Naturgefahren. Forum für Wissen, 35–42. (In German).

Frattini, P., Crosta, G., Carrara, A., Agliardi, F., 2008. Assessment of rockfall susceptibility by integrating statistical and physically-based approaches. Geomorphology 9, 419–437.

Frehner, M., Wasser, B., Schwitter, R., 2005. Nachhaltigkeit und Erfolgskontrolle im Schutzwald. Wegleitung für Pflegemassnahmen in Wäldern mit Schutzfunktion. Bern, Bundesamt für Umwelt, Wald und Landschaft (BUWAL). (In German).

Gardner, J., 1970. Rockfall a geomorphic process in high mountain terrain. Albertan Geogapher 6, 15–20.

Gardner, J., 1980. Frequency, magnitude, and spatial distribution of mountain rockfalls and rockslides in the Highwood Pass Area, Alberta, Canada. In: Coates, D.R., Vitek, J.D., (eds.), Thresholds in Geomorphology. Allen and Unwin, New York, pp. 67–295.

Gardner, J., 1983. Rockfall Frequency and Distribution. Zeitschrift für Geomorphologie 27(3), 311–324.

Gascuel, J.D., Cani-Gascuel, M.P., Desbrun, M., Leroi, E., Migron, C., 1998. Simulating landslides for natural disaster prevention. In: Arnaldi, B., Hegron, G., (eds.), Computer animation and simulation '98. Proceedings of the Eurographics Workshop, Lisbon, pp. 1–12.

Gerber, W., 1994. Beurteilung des Prozesses Steinschlag. WSL Birmensdorf, FAN – Kurs. (In German).

Gerber, W., 1998. Waldwirkungen und Steinschlag. WSL Birmensdorf, Kursunterlagen. (In German).

Gerber, W., 2001. Guideline for the approval of rockfall protection kits. Bern, Swiss Agency Environment Forests Landscape, Environment in Practice, p. 39.

Giardino, J.R., Shroder, J.F., Lawson, M.P., 1984. Tree-ring analysis of movement of a rock glacier complex on Mount Mestas, Colorado, USA. Arctic and Alpine Research 16, 299-309.

Glade, T., Lang, M., (eds.), 2003. Strategies and applications in natural hazard research using historical data. Special Volume in Natural Hazards 31(3), p. 99.

Google Earth, 2009. http://earth.google.com/ (as seen on 26 March 2009).

Goudie, A.S., Viles, H.A., 1999. The frequency and magnitude concept in relation to rock weathering. Zeitschrift für Geomorphologie 115, 175–189.

Gruber, S., Hoelzle, M., Haeberli, W., 2004. Permafrost thaw and destabilization of Alpine rock walls in the hot summer of 2003. Geophysical Research Letters 31, L13504.

Gruner, U., 2008. Climatic and meteorological influences on rockfall and rockslides ("Bergsturz"). Interpraevent 2008, Conference Proceedings, Vol. 2.

Gsteiger, P., 1993. Steinschlagschutzwald. Ein Beitrag zur Abgrenzung, Beurteilung und Bewirtschaftung. Schweizerische Zeitschrift für Forstwesen 144(2), 115–132.

Guzzetti, F., 2000. Landslide fatalities and evaluation of landslide risks in Italy. Engineering Geology 58, 89–107.

Guzzetti, F., Crosta, G., Detti, R., Agliardi, F., 2002. STONE: a computer program for the three-dimensional simulation of rock-falls. Computers and Geosciences 28, 1079-1093.

Guzzetti, F., Reichenbach, P., Wieczprek, G. F., 2003. Rockfall hazard and risk assessment in the Yosemite Valley, California, USA. Natural Hazards and Earth System Sciences 3, 491–503.

Guzzetti, F., Reichenbach, P., Ghigi, S., 2004. Rockfall hazard and risk assessment along a transportation corridor in the Nera Valley, Central Italy. Environmental Management 34(2), 191–208.

Haeberli, W., 1996. On the morphpdynamics of ice/debris-transport systems in cold mountain areas. Norsk Geografisk Tiddskrift 50, 3–9.

Haeberli, W., 1999. Hangstabilitätsprobleme im Zusammenhang mit Gletscherschwund und Permafrostdegradation im Hochgebirge. Relief, Boden, Paläoklima 14, 11–30. (In German).

Haeberli, W., Beniston, M., 1998. Climate change and its impacts on glaciers and permafrost in the Alps. Ambio 27(4), 258-265.

Hamilton, L.S., 1992. The protective role of mountain forests. GeoJournal 27, 13–22.

Hall, K., 2007. Evidence for freeze–thaw events and their implications for rock weathering in northern Canada: II. The temperature at which water freezes in rock. Earth Surface Processes and Landforms 32(2), 249–259.

Hantz, D., Vengeon, J.M., Dussauge-Peisser, C., 2003. An historical, geomechanical and probabilistic approach to rock-fall hazard assessment. Natural Hazards and Earth System Sciences 3, 693–701.

Heim, A., 1932. Bergsturz und Menschenleben. Vierteljahresschrift der Naturforschenden Gesellschaft in Zürich 77, 1–218. (In German).

Hétu, B., Gray, J.T., 2000. Effects of environmental change on scree slope development throughout the postglacial period in the Chic-Choc Mountains in the northern Gaspé Peninsula, Québec. Geomorphology 32, 335–355.

Hitz, O.M., Gärtner, H., Heinrich, I., Monbaron, M., 2008. First time application of Ash (*Fraxinus excelsior* L.) roots to determine erosion rates in mountain torrents. Catena 72, 248–258.

Hoek, E., 1987. Rockfall—a program in BASIC for the analysis of rockfall from slopes. Unpublished note, Golder Associates/University of Toronto, Canada.

Hungr, O., Evans, S.G., 1988. Engineering evaluation of fragmental rockfall hazards. In: Proceedings of the 5th International Symposium on Landslides, Lausanne, pp. 685–690.

Hungr, O., Beckie, R.D., 1998. Assessment of the hazard from rock fall on a highway. Canadian Geotechnical Journal 35(2), 409.

Hungr, O., Evans, S.G., Hazzard, J., 1999. Magnitude and frequency of rock falls along the main transportation corridors of south-western British Colombia. Canadian Geotechnical Journal 36, 224–238.

Hupp, C.R., 1984. Dendrogeomorphic evidence of debris flow frequency and magnitude at Mount Shasta, California. Environmental Geology and Water Sciences 6, 121–128.

Hupp, C.R., Osterkamp, W.R., Thornton, J.L., 1987. Dendrogeomorphic evidence and dating of recent debris flows on Mount Shasta, northern California. U.S., Geological Survey Professional Paper 1396B, 1–39.

Iesli, M., Schweingruber, F.H., 1989. Sichtbarmachen von Jahrringen für dendrochronologische Untersuchungen. Dendrochronologia 4, 145-157. (In German).

Ishikawa, M., Kurashige, Y. and Hirakawa, K., 2004. Analysis of crack movements observedn in an alpine bedrock cliff. Earth Surface Processes and Landforms 29(7), 883–891.

Jaboyedoff, M., Baillifard, F., Philippossian, F., Rouiller, J.-D., 2003. Assessing the fracture occurrence using the "Weighted fracturing density": a step towards estimating rock instability hazard. Natural Hazards and Earth System Sciences 4, 83–93.

Jacoby, G.C., 1997. Application of tree ring analysis to paleoseismology. Reviews of Geophysics 35, 109–124.

Jacoby, G.C., Sheppard, P.R., Sieh, K.E. 1988. Irregular recurrence of large earthquakes along the San Andreas Fault. Evidence from trees. Science 241, 196–199.

Jacoby, G.C., Williams, P.L., Buckley, B.M., 1992. Tree ring correlation between prehistoric landslides and abrupt tectonic events in Seattle, Washington. Science 258, 1621–1623.

Jahn, J., 1988. Entwaldung und Steinschlag. In: Proceedings International Symposium Interpraevent 1988, Graz 1, pp. 185–198.

John, K.W., Spang, R.M., 1979. Steinschläge und Felsstürze. Voraussetzungen – Mechanismen – Sicherung. Kandersteg, Tagespublikation UIC Unterausschuss K7. (In German).

Jones, C.L., Higgins, J.D., Andrew, R.D., 2000. Colorado Rockfall Simulation Program Version 4.0. Colorado Department of Transportation, Colorado Geological Survey, pp. 127.

Kalberer, M., Ammann, M., Jonsson, M., 2007. Mechanische Eigenschaften der Fichte: Experimente zur Analyse von Naturgefahren. Schweizerische Zeitschrift für Forstwesen 158, 166–175. (in German).

Kariya, Y., Sato, G., Mokudai, K., Komori, J., Ishii, M., Nishii, R., Miyazawa, Y., Tsumura, N., 2007. Rockfall hazard in the Daisekkei Valley, the northern Japanese Alps, on 11 August 2005. Landslides 4, 91–94.

Keefer, D.K., 1984: Landslides caused by earthquakes. Geological Society of America Bulletin 95, 406–421.

Keefer, D.K., 2002. Investigating landslides caused by earthquakes – historical review. Surveys in Geophysics 23, 473– 510.

Keylock, C., Domaas, U., 1999. Evaluation of topographic models of rockfall travel distance for use in hazard applications. Arctic, Antarctic, and Alpine Research 31(3), 312–20.

Kienholz, H., 1998. Landschaftsökologie II: Geomorphologie. Lecture script. Department of Geography, University of Berne. (In German).

Kirkby, M.J., Statham, I., 1975. Surface stone movement and scree formation. Journal of Geology 83, 349–62.

Kobayashi, Y., Harp, E.L., Kagawa, T., 1990. Simulation of rockfalls triggered by earthquakes. Rock Mechanics and Rock Engineering 23, 1–20.

Koo, C.Y., Chern, J.C., 1998. Modification of the DDA method for rigid block problems. International Journal of Rock Mechanics and Mining Science 35(6), 683–93.

Körner, H. J., 1980. Modelle zur Berechnung der Bergsturz- und Lawinenbewegung. In: Proceedings International Conference Interpraevent, Klagenfurt, Austria, pp. 15–55. (In German).

Kotarba, A., Strömquist, L., 1984. Transport, sorting and deposition processes of Alpine debris slope deposits in the Polish Tatra Mountains. Geografiska Annaler 66A(4), 285–294.

Kräuchi, N., Brang, P., Schönenberger, W., 2000. Forests of mountainous regions: gaps in knowledge and research needs. Forest Ecology and Management 132(1), 73–82.

Krautblatter, M., 2003. The impact of rainfall intensity and other external factors on primary and secondary rockfall (Reintal, Bavarian Alps). Thesis, University of Erlangen-Nuremberg (Germany), Department of Geography.

Krautblatter, M., Moser, M., 2006. Will we face an increase in hazardous secondary rockfall events in response to global warming in the foreseeable future? In: Price MF, editor. Global Change in Mountain Regions. Duncow: Sapiens, pp. 253–254.

Krautblatter, M., Dikau, R., 2007. Towards a uniform concept for the comparison and extrapolation of rockwall retreat and rockfall supply. Geografiska Annaler 89A(1), 21–40.

Krummenacher, B., 1995. Modellierung der Wirkungsräume von Erd- und Felsbewegungen mit Hilfe Geographischer Informationssysteme (GIS). Schweizerische Zeitschrift für Forstwesen 146, 741–761. (In German).

Krummenacher, B., Keusen, H. R., 1996. Rockfall simulation and hazard mapping based on digital terrain model (DTM). European Geologist 12, 33–35.

Kühne, R., 2005. Steinschlagsimulation in Gebirgswäldern – Validierung und Anwendung des 3D Modells ROCKYFOR in drei Testgebieten zur Analyse der Wirkung von Schutzwald. Diploma thesis. Geographisches Institut, Universität Bern, Bern. (In German).

Lafortune, M., Filion, L., Hétu, B., 1997. Dynamique d'un front forestier sur un talus d'éboulis actif en climat tempéré froid (Gaspésie, Québec). Géographie Physique et Quaternaire 51, 1–15. (In French).

LaMarche, V.C., 1961. Rate of slope erosion in the White Mountains, California. Geological Society of America Bulletin 72, 1579–1580.

LaMarche, V.C., 1966. An 800-year history of stream erosion as indicated by botanical evidence. U.S. Geological Survey Professional Paper 550D, 83–86.

Larson, P.R., 1994. The vascular cambium. Development and structure. Berlin, Springer.

Lateltin, O., 1997. Prise en compte des dangers dus aux mouvements de terrain dans le cadre des activités de l'aménagement du territoire, Recommandations, OFEFP. (In French).

Lauber, T., 1995. Bergsturz und Steinschlag im Täschgufer, Täsch. Geological advisory opinion 95-525.1, Naters, Switzerland. (In German).

Le Hir, C., 2005. Forêt et chutes de blocs: méthodologie de modélisation spatialisée du rôle de protection. Marne-La-Vallée: Univ Marne-La-Vallée, Thesis, p. 195. (In French).

Liniger, M., 2000. Computersimulation von Stein- und Blockschlägen. Felsbau 18(3), 64–68. (In German).

Luckman, B.H., 1976. Rockfalls and rockfall inventory data; some observations from the Surprise Valley, Jasper National Park, Canada. Earth Surface Processes and Landforms 1, 287–298.

Luckman, B.H., Fiske, C.J., 1995. Estimating long-term rockfall accretion rates by lichenometry. In: Slaymaker, O. (ed.), Steepland Geomorphology. Wiley, Chichester, UK, pp. 233–255.

Ludwig, K.R., Simmons, K.R., Szabo, B.J., Winograd, I.J., Landwehr, J.M., Riggs, A.C., Hoffman, R.J., 1992. Massspectrometric 230 Th-234U-238U dating of the Devils Hole calcite vein. Science 258, 284–287.

Lundström, T., Jonsson, M.J., Kalberer, M., 2007a. The root–soil system of Norway spruce subjected to turning moment: resistance as a function of rotation. Plant and Soil 300(1-2), 35–49.

Lundström, T., Jonas, T., Stöckli, V., Ammann, W., 2007b. Anchorage of mature conifers: resistive turning moment, root–soil plate geometry and root growth orientation. Tree Physiology 27, 1217–1227.

Lundström, T., Jonsson, M.J., Volkwein, A., Stoffel, M., 2009. Reactions and energy absorption of trees subject to rockfall: a detailed assessment using a new experimental method. Tree Physiology 29, 345–59.

Margreth, S., 2004. Die Wirkung des Waldes bei Lawinen. Birmensdorf, Eidgenössische Forschungsanstalt Wald Schnee Landschaft, Forum für Wissen 2004, 21–26. (In German).

Maerz, N.H., Youssef, A., Fennessey, T.W., 2005. New Risk–Consequence Rockfall Hazard Rating System for Missouri Highways Using Digital Image Analysis. Environmental & Engineering Geoscience XI(3), 229–249.

Marquínez, J., Menéndez Duarte, R., Farias, P., Jiménez Sánchez, M., 2003. Predictive GIS-based model of rockfall activity in mountain cliffs. Natural Hazards 30, 341–360.

Marzorati, S., Luzi, L., De Amicis, M., 2002. Rock falls induced by earthquakes: a statistical approach. Soil Dynamics and Earthquake Engineering 22, 65–577.

Matsuoka, N., 1990. The rate of bedrock weathering by frost action: field measurements and a predictive model. Earth Surface Processes and Landforms 15, 73–90.

Matsuoka, N., 2001. Direct observation of frost wedging in alpine bedrock. Earth Surface Processes and Landforms 26(6), 601–614.

Matsuoka, N., 2008. Frost weathering and rockwall erosion in the southeastern Swiss Alps: long term (1994-2006) observations. Geomorphology 99, 353–368.

Matsuoka, N., Sakai, H., 1999. Rockfall activity from an alpine cliff during thawing periods. Geomorphology 28, 309–328.

Matsuoka, N., Hirakawa, K., Watanabe, T., Moriwaki, K., 1997. Monitoring of Periglacial Slope Processes in the Swiss Alps: the First Two Years of Frost Shattering, Heave and Creep. Permafrost and Periglacial Processes 8, 155–177.

Matthews, J.A., Shakesby, R.A., 2004. A twentieth-century neoparaglacial rock topple on a glacier foreland, Ötztal Alps, Austria. The Holocene 14(3), 454–458.

McAuliffe, J.R., Scuderi, L.A., McFadden, L.D., 2006. Tree-ring record of hillslope erosion and valley floor dynamics: Landscape responses to climate variation during the last 400yr in the Colorado Plateau, northeastern Arizona. Global and Planetary Change 50, 184–201.

McCarroll, D., Shakesby, R.A., Matthews, J.S., 1998. Spatial and temporal patterns of Late Holocene rockfall activity on a Norwegian talus slope: lichenometry and simulation-modelling approach. Arctic and Alpine Res. 30, 51–60.

McCarroll, D., Shakesby, R.A., Matthews, J. 2001. Enhanced Rockfall Activity during the Little Ice Age: Further lichenometric evidence from a Norwegian talus. Permafrost and Periglacial Processes 12, 157–164.

Meissl, G., 1998. Modellierung der Reichweite von Felsstürzen. Fallbeispiele zur GIS-gestützten Gefahrenbeurteilung aus dem Beierischen und Tiroler Alpenraum. Innsbrucker Geographische Studien 28, Institut für Geographie, Universität Innsbruck, pp. 249. (In German).

Meissl, G., 2001. Modelling the runout distances of rockfalls using a geographic information system. Zeitschrift für Geomorphologie 125, 129–137.

Menéndez Duarte, R., Marquínez, J., 2002. The influence of environmental and lithologic factors on rockfall at a regional scale: an evaluation using GIS. Geomorphology 43, 117–136.

Moore, D.P., Mathews, W.H., 1978. The Rubble Creek landslide, southwestern British Columbia. Canadian Journal of Earth Sciences 15, 1039–1052.

Motta, R., Haudemand, J.C., 2000. Protective forests and silvicultural stability. An example of planning in the Aosta valley. Mountain Resort Development 20, 74–81.

Mundo, I.A., Barrera, M. D., Roig, F. A., 2007. Testing the utility of *Nothofagus pumilio* for dating a snow avalanche in Tierra del Fuego, Argentina. Dendrochronologia 25, 19–28, 2007.

Muntán, E., Andreu, L., Oller, P., Gutiérrez, E., Martínez, P., 2004. Dendrochronological study of the Canal del Roc Roig avalanche path: first results of the Aludex project in the Pyrenees. Annals of Glaciology 38, 173–179.

Muntán, E., Oller, P., García, C., Martí, G., García, A., Gutiérrez, E., 2008. Reconstructing snow avalanches in the Southeastern Pyrenees. Natural Hazards and Earth System Sciences. in review.

Nagy, N.E., Franceschi, V.R., Solheim, H., Krekling, T., Christiansen, E., 2000. Wound induced traumatic resin duct formation in stems of Norway spruce (Pinaceae): anatomy and cytochemical traits. American Journal of Botany 87, 302–313.

Nishiizumi, K., Kohl, C.P., Arnold, J.R., Dorn, R., Klein, J., Fink, D., Middelton, R., Lal, D., 1993. Role of *in situ* cosmogenic nuclides 10Be and 26Al in the study of diverse geomorphic processes. Earth Surface Processes and Landforms 18, 407–425.

Nyberg, R., 1991. Geomorphic processes at snowpatch sites in the Abisko Mountains, northern Sweden. Zeitschrift für Geomorphologie 35(3), 321–343.

O'Hara, K.L., 2006. Multiaged forest stands for protection forests: concepts and applications. Forest Snow and Landscape Research 80(1), 45–56.

Okura, Y., Kitahara, H., Sammori, T., Kawanami, A., 2000a. The effects of rockfall volume on runout distance. Engineering Geology 58(2), 109–124.

Okura, Y., Kitahara, H., Sammori, T., 2000b. Fluidization in dry landslides. Engineering Geology 56(3), 347–60.

6. Bibliography of Chapter A - 63

Omura, H., Marumo, Y., 1988. An experimental study of the fence effects of protection forests on the interception of shallow mass movement. Mitteilungen der Forstlichen Bundes-Versuchsanstalt Mariabrunn Wien 159, 139–147.

Ott, E., Frehner, M., Frey, H., Lüscher, P., 1997. Gebirgsnadelwälder: Ein Praxisorientierter Leitfaden für eine Standortgerechte Waldbehandlung. Paul Haupt Verlag, Bern. (In German).

Paronuzzi, P., Artini, E., 1999. Un nuovo programma in ambiente Windows per la modellazione della caduta massi. Geologia Tecnica e Ambientale 1/99, 13–24. (In Italian).

Peckover, F.L., Kerr, J.W.G., 1977. Treatment and maintenance of rock slopes on transportation routes. Canadian Geotechnical Journal 14, 458–507.

Peltola, H., Kellomäki, S., Hassinen, A., Granader, M., 2000. Mechanical stability of Scots pine, Norway spruce and birch: an analysis of tree-pulling experiments in Finland. Forest Ecology and Management 135, 143–153.

Penck, W., 1924. Die morphologische Analyse. J. Engelshorn Verlag. Stuttgart. (In German).

Perret, S., Dolf, F., Kienholz, H., 2004. Rockfalls into forests: Analysis and simulation of rockfall trajectories — considerations with respect to mountainous forests in Switzerland. Landslides 1, 123–130.

Perret, S., Stoffel, M., Kienholz, H., 2006. Spatial and temporal rockfall activity in a forest stand in the Swiss Prealps—A dendrogeomorphological case study. Geomorphology 74, 219–231.

Pfeiffer, T.J., Bowen, T., 1989. Computer simulation of rockfalls. Bulletin of the Association of Engineering Geologists 26(1), 135–146.

Pfeiffer, T.J., Higgins, J.D., Schultz, R., Andrew, R.D., 1991. Colorado Rockfall Simulation Program Users Manual for Version 2.1. Colorado Department of Transformation, Denver, pp. 127.

Pierson, L.A., Davis, S.A., Van Vickle, R., 1990. Rockfall hazard rating system. Implementation Manual. Washington DC: Federal Highway Administration, U.S. Department of Transportation, Report, FHWA-OR-EG-90-01.

Piteau, D.R., Clayton, R., 1976. Computer Rockfall Model. In: Proceedings of the Meeting on Rockfall Dynamics and Protective Works Effectiveness, Bergamo, Italy, ISMES Publication No. 90, pp. 123–125.

Porter, S.C., Orombelli, G., 1981. Alpine rockfall hazards. American Scientist 69, 67-75.

Raetzo, H., Lateltin, O., Bollinger, D., Tripet, J. P., 2002. Hazard assessment in Switzerland – codes of practice for mass movements. Bulletin of Engineering Geology and the Environment 61, 263-268.

Rapp, A., 1960. Recent developments in the mountain slopes in Kärkevagge and surroundings, northern Scandinavia. Geografiska Annaler 42, 1–158.

Rickli, C., Zimmerli, P., Böll, A., 2001. Effects of Vegetation on Shallow Landslides - an Analysis of the Events of August 1997 in Sachseln, Switzerland. Kühne, M. (ed.), Proceedings of the International Conference on Landslides 2001, Davos, 575 - 584.

Rickli, C., Graf, F., Gerber, W., Frei, M., Böll, A., 2004. Der Wald und seine Bedeutung bei Naturgefahren geologischen Ursprungs. In: Eidgenössische Forschungsanstalt WSL (ed.), Schutzwald und Naturgefahren. Forum für Wissen 2004. Bruhin AG, Freienbach, pp. 27–34.

Rinntech, 2009. http://www.rinntech.com/ (see LINTAB and TSAP-Win, as seen on 12 March 2009).

Ritchie, A.M., 1963. Evaluation of rockfall and its control. Highway Research Board Record 17, 13–28.

Ritter, D.F., Kochel, G.R., Miller, J.R., 2002. Process geomorphology. McGraw-Hill, New York (4th edition).

Rodríguez, C.E., Bommer, J., Chandler, R.J., 1999. Earthquake-induced landslides: 1980–1997. Soil Dynamics and Earthquake Engineering 18, 325–346.

Romana, M., 1988. Practice of SMR classification for slope appraisal. In: Proceedings of the 5th International Symposium on Landslides, Lausanne (10–15 Jul 1988). Rotterdam, Balkema, pp. 1227–1229.

Rosser, N.J., Petley, D.N., Lim, M., Dunning, S.A., Allison, R.J., 2005. Terrestrial laser scanning for monitoring the process of hard rock costal cliff erosion. Quarterly Journal of Engineering Geology and Hydrogeology 38, 363–375.

Rouiller, J.-D., Jaboyedoff, M., Marro, C., Phlippossian, F., 1997. Matterock: méthodologie d'étude d'instabilités de falaise et d'appréciation du danger. Société suisse de mécanique des sols et des roches 135, 13–16. (In French).

Rovera, G., Robert, Y., Coubat, M., 1999. L'action des processus périglaciaires dans les badlands marneux des Alpes du Sud: l'exemple du bassinversant du Saignon. Environnements périglaciaires (Association Française du Périglaciaire) 6, 41–52. (In French).

Rubiales, J.M., Bodoque, J.M., Ballesteros, J.A., Díez-Herrero, A., 2008. Response of *Pinus sylvestris* roots to sheet-erosion exposure: An anatomical approach. Natural Hazards and Earth System Sciences 8, 223–231.

Rupé, C., 1991. Étude de peuplement forestier de protection du canton de la Saucisse, Saint-Martin le Vinoux. Internal report Cemagref Grenoble, pp. 35. (In French).

Ruppen, P.J., Imseng, G., Imseng, W., 1979. Saaser Chronik 1200–1979. Rotten-Verlag, Brig, Valais, Switzerland. (In German).

Sachs, T., 1991. Pattern formation in plant tissue. Cambridge University Press, Cambridge.

Sandersen, F., Bakkehoi, S., Hestnes, E., Lied, K., 1996. The influence of meteorological factors on the initiation of debris flows, rockfalls, rockslides and rockmass stability. In: Senneset, K., (ed.), 7th International Symposium on Landslides, Rotterdam.

Sandersen, F., Bakkehoi, S., Hestens, E., Lied, K., 1997. The influence of meteorological factors on the initiation of debris flows, rockfalls, rockslides and rockmass stability. Publikasjon - Norges Geotekniske Institutt 201, 97–114.

Sass, O., 1998. Die Steuerung von Steinschlagmenge durch Mikroklima, Gesteinsfeuchte und Gesteinseigenschaften im westlichen Karwendelgebirge. Münchner Geographische Abhandlungen Reihe B29, 347–359. (In German).

Sass, O., 2005. Spatial patterns of rockfall intensity in the northern Alps. Zeitschrift für Geomorphologie 138, 51–65.

Schiermeier, Q., 2003. Alpine thaw breaks ice over permafrost's role. Nature 424, 712.

Schweigl, J., Ferretti, C., Nossing, L., 2003. Geotechnical characterization and rockfall simulation of a slope: a practical case study from South Tyrol (Italy). Engineering Geology 67, 281–296.

Schweingruber, F.H., 1983. Der Jahrring. Standort, Methodik, Zeit und Klima in der Dendrochronologie. Paul Haupt, Bern, Stuttgart, Wien. (In German).

Schweingruber, F.H., 1996. Tree Rings and Environment. Dendroecology. Paul Haupt, Bern, Stuttgart, Wien. (In German).

Schweingruber, F.H., 2001. Dendroökologische Holzanatomie. Paul Haupt, Bern, Stuttgart, Wien. (In German).

Schwitter, R., 1998. Zusammenfassung und Schlussfolgerungen. In: Proc 14. Arbeitstagung Schweizerischen Gebirgswaldpflegegruppe & FAN, Grafenort/Engelberg, unpublished report. pp. 1–5. (In German).

Scioldo, G., 1991. La statistica Robust nella simulazione del rotolamento massi. In: Proceedings Meeting La meccanica delle rocce a piccola profondità, Torino, Italy, pp. 319–323. (In Italian).

Scuderi, L., McFadden, L., McAuliffe, J., 2008. Dendrogeomorphically derived slope and stream response to decadal and centennial scale climate variability: Implications for downstream sedimentation. Natural Hazards and Earth System Sciences 8, 869–880.

Selby, M.J., 1993. Hillslope Materials and Processes. 2nd Edition. Oxford University Press.

Sepulveda, S.A., Murphy, W., Petley, D.N., 2004. The role of topographic amplification on the generation of earthquake-induced rock slope failures. In: Lacerda, W., Erlich, M., Fontoura, S.A.B., Sayao, A.S.F., (eds), Landslides: Evaluation and Stabilisation, Proceedings of the 9th International Sympo-

6. Bibliography of Chapter A

sium on Landslides. Balkema, Rio de Janeiro, p. 311–315.

Shigo, A.L., 1984. Compartmentalization: a conceptual framework for understanding how trees grow and defend themselves. Annual Review of Phytopathology 22, 189–214.

Shroder, J.F., 1978. Dendrogeomorphological analysis of mass movement on Table Cliffs Plateau, Utah. Quaternary Research 9, 168–185.

Shroder, J.F., 1980. Dendrogeomorphology: review and new techniques of tree-ring dating, Progress in Physical Geography 4, 161–188.

Shroder, J.F., Butler, D.R., 1987. Tree-ring analysis in the earth sciences. In: Jacoby, G.C., Hornbeck, J.W., (eds.), Proceedings of the International Symposium "Ecological Aspects of Tree-Ring Analysis", Lamont Doherty Geological Observatory, 186–212.

Stevens, W., 1998. RocFall: a tool for probabilistic analysis, design of remedial measures and prediction of rockfalls. M.A.Sc. Thesis, Department of Civil Engineering, University of Toronto, Ontario, Canada, pp. 105.

Stoffel, M., 2005a. Spatio-temporal analysis of rockfall activity into forests – results from tree-ring and tree analysis. PhD thesis. Department of Geosciences, Geography, University of Fribourg. GeoFocus 12, 1–188.

Stoffel, M., 2005b. Assessing the vertical distribution and visibility of scars in trees. Schweizerische Zeitschrift für Forstwesen 156(6), 195–199.

Stoffel, M., 2006. A Review of Studies Dealing with Tree Rings and Rockfall Activity: The Role of Dendrogeomorphology in Natural Hazard Research. Natural Hazards 39, 51–70.

Stoffel, M., 2008. Dating past geomorphic processes with tangential rows of traumatic resin ducts. Dendrochronologia 26(1), 53–60.

Stoffel, M., Perret, S., 2006. Reconstructing past rockfall activity with tree rings: some methodological considerations. Dendrochronologia 24(1), 1-15.

Stoffel, M., Bollschweiler, M. 2008. Tree-ring analysis in natural hazards research – an overview. Natural Hazards and Earth System Sciences 8, 187–202.

Stoffel, M., Bollschweiler, M. 2009. What tree rings can tell about earth-surface processes. Teaching the principles of dendrogeomorphology. Geography Compass 3, 1013–1037.

Stoffel, M., Hitz, O.M., 2008. Snow avalanche and rockfall impacts leave different anatomical signatures in tree rings of *Larix decidua*. Tree Physiology 28(11), 1713–1720.

Stoffel, M., Lièvre, I., Monbaron, M., Perret, S., 2005a. Seasonal timing of rockfall activity on a forested slope at Täschgufer (Valais, Swiss Alps) – a dendrochronological approach. Zeitschrift für Geomorphologie 49(1), 89–106.

Stoffel, M., Schneuwly, D., Bollschweiler, M., Lièvre, I. , Delaloye, R., Myint, M., Monbaron, M., 2005b. Analyzing rockfall activity (1600-2002) in a protection forest – a case study using dendrogeomorphology. Geomorphology 68(3–4), 224–241.

Stoffel, M., Lièvre, I., Conus, D., Grichting, M.A., Raetzo, H., Gärtner, H.W., Monbaron, M., 2005c. 400 years of debris flow activity and triggering weather conditions: Ritigraben VS, Switzerland. Arctic, Antarctic and Alpine Research 37(3), 387–395.

Stoffel, M., Wehrli, A., Kühne, R., Dorren, L.K.A., Perret, S., Kienholz, H., 2006a. Assessing the protective effect of mountain forests against rockfall using a 3D simulation model. Forest Ecology and Management 225, 113-122.

Stoffel, M., Bollschweiler, M., Hassler, G.R., 2006b. Differentiating events on a cone influenced by debris-flow and snow avalanche activity – a dendrogeomorphological approach, Earth Surface Processes and Landforms 31(11), 1424–1437.

Stoffel, M., Bollschweiler, M., Leutwiler, A., Aeby, P., 2008a. Large debris-flow events and overbank sedimentation in the Illgraben torrent (Valais Alps, Switzerland). Open Geology Journal 2, 18–29.

Stoffel, M., Conus, D., Grichting, M.A., Lièvre, I., Maître, G., 2008b. Unraveling the patterns of late Holocene debris-flow activity on a cone in the central Swiss Alps: chronology, environment and implications for the future. Global and Planetary Change 60, 222–234 .

Stokes, A., Salin, F., Kokutse, A.D., Berthier, S., Jeannin, H., Mochan, S., Dorren, L., Kokutse, N., Abd.Ghani, M., Fourcaud, T., 2005. Mechanical resistance of different tree species to rockfall in the French Alps. Plant and Soil 278, 107–117.

Strunk, H., 1989. Dendrogeomorphology of debris flows, Dendrochronologia 7, 15–25.

Strunk, H., 1991. Frequency distribution of debris flow in the Alps since the "Little Ice Age". Zeitschrift für Geomorphologie 83, 71–81.

Strunk, H., 1997. Dating of geomorphological processes using dendrogeomorphological methods. Catena 31, 137–151.

Swisstopo, 2009. http://www.swisstopo.admin.ch/ (as seeon on 26 March 2009).

Thormann, J.-J., Schwitter, R., 2004. Nachhaltigkeit im Schutzwald (NaiS), Nachhaltige Schutzwaldpflege und Waldbauliche Erfolgskontrolle. International Congress Interpraevent, Conference Proceedings (1III), Riva del Garda, pp. 331-342.

Tianchi, L., 1983. A mathematical model for predicting the extent of a major rockfall. Zeitschrift für Geomorphologie 27(4), 473–82.

Timell, T.E., 1986. Compression wood in Gymnosperms. Springer, Berlin.

Toppe, R., 1987. Terrain models – a tool for natural hazard mapping. In: Salm, B., Gubler, H., (eds.), Avalanche formation, movement and effects. IAHS Publication no. 162, 629–38.

Van Dijke, J.J., Van Westen, C.J., 1990. Rockfall hazard, a geomorphological application of neighbourhood analysis with ILWIS. ITC Journal 1, 40–44.

Varnes, D.J., 1978. Slope movements: types and processes. In: Schuster, R.L., Krizek, R.J., (eds.), Landslide analysis and control. Transportation Research Board, Special Report 176, Washington D.C., pp. 11–33.

Walder, J.S., Hallet, B., 1986. The physical basis of frost weathering: toward a more fundamental and unified perspective. Arctic and Alpine Research 18, 27–32.

Wasser, B., Frehner, M., 1996. Minimale Pflegesmassnahmen für Wälder mit Schutzfunktion. Wegleitung, Bundesamt für Umwelt, Wald und Landschaft (BUWAL), Bern, p. 122. (In German).

Wehrli, A., Dorren, L.K.A., Berger, F., Zingg, A., Schönenberger, W., Brang, P., 2006. Modelling the long-term impacts of forest dynamics on the protective effect against rockfall. Forest Snow and Landscape Research 80(1), 57–76.

Wentworth, C.K., 1922. A scale of grade and class terms for clastic sediments. The Journal of Geology 30, 377–392.

Whalley, W.B., 1974. The mechanics of high-magnitude low-frequency rock failure. Reading Geographical Papers, 27.

Whalley, W.B., 1984. Rockfalls. In: Brundsden, D., Prior, D.B., (eds.), Slope instability. London, Wiley, pp. 217–256.

Woltjer, M., Rammer, W., Brauner, M., Seidl, R., Mohren, G.M.J., Lexer, M.J., 2008. Coupling a 3D patch model and a rockfall module to assess rockfall protection in mountain forests. Journal of Environmental Management 87, 373–388.

Wu, S.-S. 1985. Rockfall evaluation by computer simulation. Transportation Research Record 1031, 1–5.

Zinggeler, A., 1989. Steinschlagsimulation in Gebirgswäldern. Modellierung der relevanten Teilprozesse. Diploma thesis. Geographisches Institut, Universität Bern, Bern. (In German).

✦✦✦✦✦

Chapter B

Fundamental Research

published in Tree Physiology 2009, Vol. 29, 281–289

Formation and spread of callus tissue and tangential rows of resin ducts in *Larix decidua* and *Picea abies* following rockfall impacts

Dominique M. Schneuwly [1,2], *Markus Stoffel* [1,3,4] *and Michelle Bollschweiler* [1,3,4]

[1] *Department of Geosciences, Geography, University of Fribourg,*
  Chemin du Musée 4, CH-1700 Fribourg, Switzerland

[2] *Corresponding author (dominique.schneuwly@unifr.ch)*

[3] *Department of Environmental Sciences, University of Geneva,*
  Chemin de Drize 7, CH-1227 Carouge-Geneva, Switzerland

[4] *Laboratory of Dendrogeomorphology, Institute of Geological Sciences, University of Berne,*
  Baltzerstrasse 1-3, CH-3012 Berne, Switzerland

Summary

After mechanical wounding, callus tissue and tangential rows of traumatic resin ducts (TRDs) are formed in many conifer species. This reaction can be used to date past events of geomorphic processes such as rockfall, debris flow and snow avalanches. However, only few points are known about the tangential spread or the timing of callus tissue and TRD formation after wounding. We analyzed 19 *Larix decidua* Mill. (European larch) and eight *Picea abies* (L.) Karst. (Norway spruce) trees that were severely damaged by rockfall activity, resulting in a total of 111 injuries. Callus tissue appeared sparsely on the cross sections and was detected on only 4.2% of the *L. decidua* samples and 3.6% of the *P. abies* samples. In contrast, TRDs were present on all cross sections following wounding and were visible on more than one-third (34% in *L. decidua* and 36.4% in *P. abies*) of the circumference where the cambium was not destroyed by the rockfall impact. We observe different reactions in the trees depending on the seasonal timing of wounding. The tangential spread of callus tissue and TRDs was more important if the injury occurred during the growth period than during the dormant season, with the difference between seasons being more pronounced for callus tissue formation than for TRD formation. We observed an intra-annual radial migration of TRDs with increasing tangential distance from the wound in 73.2% of the *L. decidua* samples and 96.6% of the *P. abies* samples. The persistence of TRD formation in the years following wounding showed that only *L. decidua* trees produced TRDs 2 years after wounding (10.5%), whereas *P. abies* trees produced TRDs 5 years after wounding (> 50%).

Keywords

dendrogeomorphology, European larch, injury, Norway spruce, tangential extension, traumatic resin ducts.

Introduction

Rockfall is a common mass movement process in alpine environments. After Berger et al. (2002), it is defined as the free falling, bouncing or rolling of individual or a few rocks and boulders, with volumes involved generally being < 5 m3. The collision of a boulder with a tree can cause various damages, such as wounding, tilting and breaking of the stem. Tree responses to collisions include formation of callus tissue (Schweingruber 2001), eccentric growth or the formation of reaction wood (Timell 1986, Braam et al. 1987), abrupt changes in growth (Strunk 1997, Friedman et al. 2005), and multiple leader formation (Mattheck 1996, Schweingruber 1996). In addition, various conifer spe-cies react to mechanical damage with the formation of tan-gential rows of traumatic resin ducts (TRDs, LePage and Be´gin 1996, Stoffel and Perret 2006, Bollschweiler et al. 2008). The presence and distribution of TRDs have been used to date past geomorphic events (Bollschweiler 2007, Schneuwly and Stoffel 2008a, 2008b, Stoffel 2008). Never-theless, there is still uncertainty about the exact timing and distribution of TRD formation after mechanical impact on the stem. In contrast to widely scattered resin ducts that appear as a common feature in stems of the genus *Larix* (Bannan 1936), TRD formation operates as a defense mech-anism that compartmentalizes the wood after tree damage (Thomson and Sifton 1925, Shigo 1984, Hudgins et al. 2004). If the injury occurs during a period of cambial activity, the TRDs differentiate from the cambium with a delay of 4–28 days (Nagy et al. 2000, Franceschi et al. 2002, Luchi et al. 2005). Damage occurring outside the growth period causes formation of TRDs during the following growing season (Fahn et al. 1979). The TRDs normally appear in tangential rows close to the wound where the cambium has been partially destroyed, and their arrangement becomes more dispersed away from the wound (Bannan 1936). Fahn et al. (1979) had concluded that the number and size of ducts decreases with distance from the injury. There are few published quantitative data on the tangential extent of TRDs (Moore 1978, Bollschweiler et al. 2008, Stoffel and Hitz 2008). The lateral and axial spread of TRDs seems to depend not only on the distance to the injury, but also on the season when wounding occurred. The most pronounced appearance of TRDs after tree damage is observed when the injury occurs during the growing season, whereas wounding during the dormant period results in a weaker spread of TRD in the next growing season (Bannan 1936). Another unknown factor is the intra-annual shift of TRDs within a growth ring with increasing distance from the wound. Bannan (1936) had described the phenomenon of a delayed onset of TRD formation, but had only mentioned an axial shift of ducts. Bollschweiler et al. (2008) and Stoffel and Hitz (2008) had observed, in addition to the axial shift, a radial migration of TRD with increasing distance from the injury. The same authors had also investigated the persistence of TRD formation, but these studies were on juvenile plants or trees that had been disturbed by debris-flow activity.

As a result, there is only limited knowledge on the tangential and intra-annual occurrence and the persistence of TRD formation in trees wounded by falling rocks and boulders. Similarly, there have been no studies on the anatomic responses to natural wounding in *Picea abies* (L.) Karst. (Norway spruce). To improve both the nondestructive investigation of naturally injured trees with increment cores and the accurate dendrogeomorphologic dating of past rockfall activity, it is crucial to improve our understanding of the axial and intra-annual behavior of TRD formation.

The aims of this study were to: (1) investigate and quantify the formation and radial extension of callus tissue and TRDs; (2) determine the influence of seasonality on the formation of callus tissue and TRDs; (3) analyze the intra-annual shift of TRDs with increasing tangential distance from the wound; and (4) assess the persistence of TRD for-mation in the years following wounding.

Materials and methods

Study site

The study site is situated 5 km north of Saas Fee close to the village of Saas Balen (46°09'06" N and 7°55'27" E), Valais, Switzerland. The forest stand is located between 1390 and 1610 m a.s.l. with a mean slope of 36°. Within the forest stand, a thin layer of quaternary talus and morainic deposits covers the bedrock that comprises micaceous schists belonging to the Penninic crystalline layers. The forest consists mainly of *Larix decidua* Mill. (European larch, 90%), some *P. abies* (10%), and a few *Pinus cembra* L. (< 1%) trees. Phenological data from a nearby site (10 km to the east; Müller 1980) indicate that the growth period is initiated around mid-May, latewood formation starts in mid-July and tree-ring growth ceases in mid-October. Mean annual temperature (1987–2007) at the study site is 4.5 °C with a mean annual rainfall of 620 mm (SMI (Swiss Meteorological Institute) 2008).

Vegetation on the study site is strongly influenced by rockfall activity that regularly occurs on the slope (Schneuwly and Stoffel 2008a, 2008b), mainly outside the growing season. The volume of the falling boulders does not normally exceed 1 m^3, and the anthropogenic influence is negligible in the forest stand.

Field work and sample preparation

At the study site, no geomorphic processes other than rockfall could be identified, and there were no signs of avalanche paths or debris-flow channels on the study slope (Figure 1A). Based on the analysis of processes present on the site, severely injured trees with multiple rockfall injuries were selected for this study. For the study trees, we recorded tree species, tree height and tree diameter at breast height (DBH). All visible defects in each tree's morphology were recorded and pictures of each individual were taken. The maximum extension, width, height and orientation of each injury were noted before a picture was taken of every wound (Figure 1B). Finally, one cross section was sawn from each wound at its maximum lateral extension, resulting in a dataset of 111 injuries (82 for *L. decidua*, 29 for *P. abies*).

Laboratory analysis

In the laboratory, tree reaction after wounding was analyzed. Samples were first air-dried and polished with sand-paper up to 400 grit. In addition, a selected number of samples were used for detailed microscopic examination, and micro-cuts were prepared from the wounded area and the adjacent tissue, as described by Schweingruber (1978) and Clark (1981). The width (in degrees), the year and the intra-annual position of the injury were assessed. We noted the earliest evidence of a reaction, such as changes in cell structure (Stoffel and Hitz 2008), formation of callus tissue and appearance of TRDs. To determine the intra-annual timing of wounding, reactions in the growth ring were subdivided as described by Stoffel et al. (2005) in Figure 2B into: the first-formed cell layer (in which TRD formation usually occurs as the result of an injury during the dormant season, D), early, middle and late earlywood (EE, ME and LE, respectively), and early and late latewood (EL and LL, respectively).

Thereafter, the nature of the TRDs was determined and subdivided into three classes: in Class 1, TRDs were formed as extremely compact and continuous rows; in Class 2, TRDs were very compact but not formed in completely continuous rows; in Class 3, TRDs were tangentially aligned but with clearly observable gaps between single ducts. The tangential extension of each TRD class was assessed on both sides of the injury by measuring the distance between the wound's boundary and the farthest point at which TRDs of the different classes occurred. Following Bollschweiler et al. (2008), these distances are given as a percentage of the ring's total circumference, excluding that portion where the cambium had been destroyed (Figure 2A). In addition, the seasonality of TRD occurrence was noted for each class. During the analysis of TRD formation, we focused on the first year of appearance but also noted the presence of TRDs in subsequent years and for how many years they were present. In a last analytic step, the radial shift in TRDs with increasing

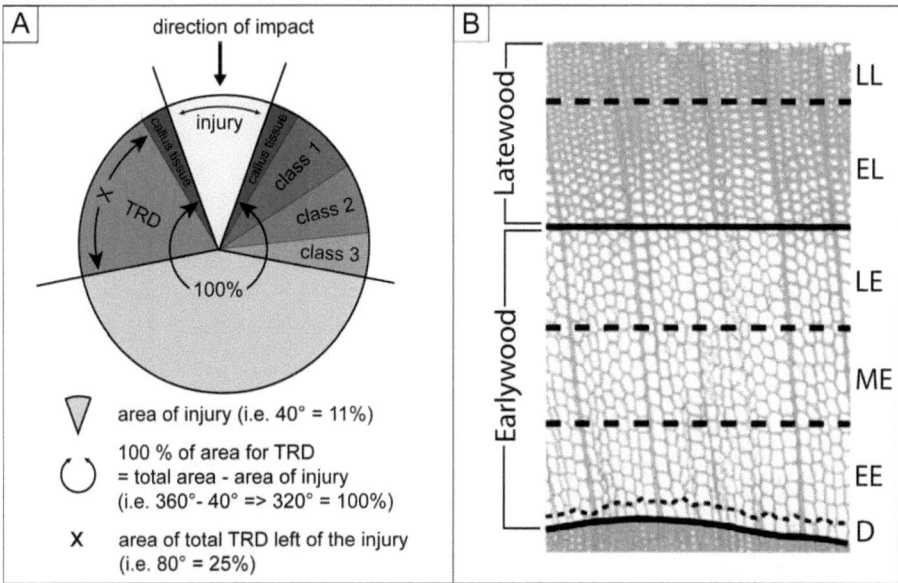

Figure 1. (A) General view of the rockfall slope, with the investigated forest stand indicated by an arrow. (B) Typical rockfall injury, notice the wood chipped off by the impact.

Figure 2. (A) Schematic illustration of the tangential TRD analysis. (B) Intra-annual subdivisions of a growth ring into: dormancy (D), early earlywood (EE), middle earlywood (ME), late earlywood (LE), early latewood (EL) and late latewood (LL).

distance from the wound was assessed. We noted the maximum extent (in mm) of TRDs on both sides of the wound and noted the intra-annual position (Figure 2B) of TRDs next to the injury and at the outermost position. We then counted the number of seasonality classes (D, EE, ME, LE, EL and LL) between the injury and the intra-annual position of the outermost TRDs. These shifting values, summarized in categories, were finally related to the mean distance between the injury and the outermost TRDs measured previously.

Results

General aspects and seasonal distribution

We investigated 27 trees that were severely injured by rockfall (Table 1), comprising a total of 111 injuries.

	Total	*L. decidua*	*P. abies*
Number of trees	27	19	8
Number of injuries	111	82	29

Table 1. Number of L. decidua and P. abies trees and injuries sampled.

The distribution of tree ages and DBH of the individuals at the time of wounding as well as an overview on injury widths is provided in Table 2. Tree age at the moment of wounding averaged 11.3 years for the *L. decidua* trees and 15.2 years for the *P. abies* trees. The age distribution showed a maximum number of injuries when the trees were 6–8 (Larix) and 12–14 (Picea) years old. The mean DBH at the time of injury averaged 45 mm for *L. decidua* (peak at 21–30 mm) and 47.2 mm for *P. abies* (peak at 41–50 mm). Mean injury widths were similar between species, with 85.7 for *L. decidua* and 92.7 for *P. abies*. The distribution of injury widths reached a maximum at 41–60% (*L. decidua*) and 61–80% (*P. abies*). Thus, tree age and DBH at the time of injury as well as injury widths were comparable between the study species.

The distribution of the intra-annual position of the earliest response after wounding revealed that most reactions started at the very beginning of the growth ring. In 76.6% of all injuries, no cell rows were formed before the formation of callus tissue or TRDs, implying that most of the impacts were induced outside the growth period between mid-October and mid-May. Although 19.8% of tree reactions

Age classes	*L. decidua*	*P. abies*	DBH classes	*L. decidua*	*P. abies*	Injury width classes	*L. decidua*	*P. abies*
3–5	14.6	6.9	≤ 20	15.9	10.3	≤ 20	2.4	*0*
6–8	**30.5**	10.3	21–30	**19.5**	13.8	21–40	9.8	10.3
9–11	20.7	13.8	31–40	13.4	6.9	41–60	**22**	20.7
12–14	7.3	**24.1**	41–50	17.1	**31**	61–80	14.6	**24.1**
15–17	9.8	17.2	51–60	11	13.8	81–100	20.7	10.3
18–20	4.9	6.9	61–70	3.7	13.8	101–120	13.4	10.3
21–23	4.9	10.3	71–80	9.8	*0*	121–140	11	10.3
24–26	2.4	*0*	81–90	6.1	10.3	141–160	2.4	3.4
27–30	2.4	3.4	91–100	*1.2*	*0*	161–180	*1.2*	3.4
> 30	2.4	6.9	> 100	2.4	*0*	> 180	2.4	6.9
Overall statistics (years)			Overall statistics (mm)			Overall statistics (°)		
Mean	11.3	15.2	Mean	45	47.2	Mean	85.7	92.7
Minimum	*3*	*3*	Minimum	*12*	*15*	Minimum	*13*	*23*
Maximum	**30**	**32**	Maximum	**150**	**89**	Maximum	**315**	**240**
SD	6.86	7.33	SD	25.4	20.3	SD	47	53.5

Table 2. Distribution of age (years) and DBH (mm) at time of wounding and distribution of injury widths (°) in the L. decidua and P. abies trees (%) studied. Minimum values are given in italics, maximum values are highlighted in bold.

were observed during the earlywood period (mid-May to mid-July), only 3.6% were initiated during the formation of latewood cell layers (mid-July to mid-October) (Table 3).

Season	D	EE	ME	LE	EL	LL	Total
Proportion (%)	76.6	9	9	1.8	1.8	1.8	100

Table 3. Season of wounding. Intra-annual position of the earliest response after wounding (dormant season (D), early (EE), middle (ME) and late earlywood (LE), early (EL) and late (LL) latewood). The distribution reveals a strong peak in rockfall activity in the dormant season which lasts from mid-October to mid-May.

Arrangement of callus tissue and TRD in the year of injury

In general, TRD formation represented the most common growth feature following wounding. Following an injuring event, TRDs appeared on all samples without exception. In contrast, callus tissue was present on only 45.1% of L. decidua samples and 41.4% of P. abies samples.

Table 4 shows that the radial spread of TRDs was iden-tified on 34% (L. decidua) and 36.4% (P. abies) of the ring's circumference that was remaining vital after impact. Callus tissue was, in contrast, detected only on 4.2% (L. decidua) and 3.6% (P. abies) of the circumference where the cam-bium was not destroyed. There was less difference between species in either feature (Figures 3 and 4).

Influence of the seasonal timing of wounding on callus tissue and TRD formation

Eighty-five injuries were inflicted during the dormant season and only 26 injuries occurred during the growth period. The formation of callus tissue and TRDs differed between trees injured during the dormant season and those damaged during the growing season. As illustrated in Table 5, the formation of callus tissue was much more pronounced, by a factor of 2.25, in the samples that were injured during the growth period. Similarly, the tangential extension of TRDs was more pronounced, by a factor of 1.25, if the impact occurred during the growing season. Thus, callus tissue formation depended much more on the seasonality of the event than TRD formation

Injury	No.	TRD (%)	Callus tissue (%)
During dormant season	85	32.7	18.8
During growth period	26	40.7	42.3

Table 5. Reaction to wounding in L. decidua and P. abies depends on season of wounding.

	Injury (°)	Callus tissue (%*)	TRD sum (%*)	TRD I (%*)	TRD II (%*)	TRD III (%*)
L. decidua						
Mean	85.7	4.2	34	22	6.4	5.6
SD	47	7.4	32.4	20	6.8	5.6
P. abies						
Mean	92.7	3.6	36.4	16.2	11.8	8.4
SD	53.5	6.4	44.8	22	14.2	8.6

Table 4. Tree reaction to wounding assessed as injury width (), spread of callus tissue and classes of TRD in the L. decidua and P. abies samples. Abbreviation: (%*) represents that part of the tree's circumference remaining vital after the impact.

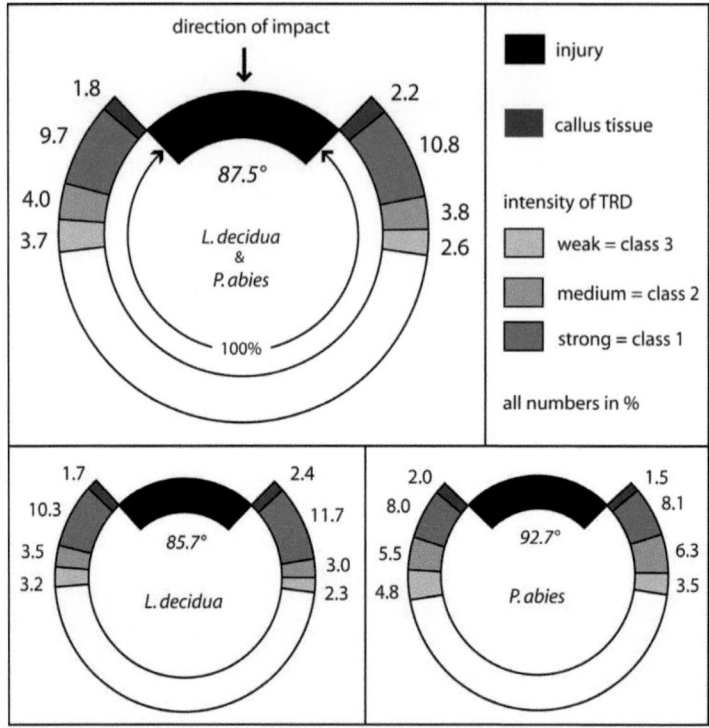

Figure 3. Distribution of callus tissue and intensity of TRD (classes 1–3) in the year of wounding in L. decidua and P. abies trees.

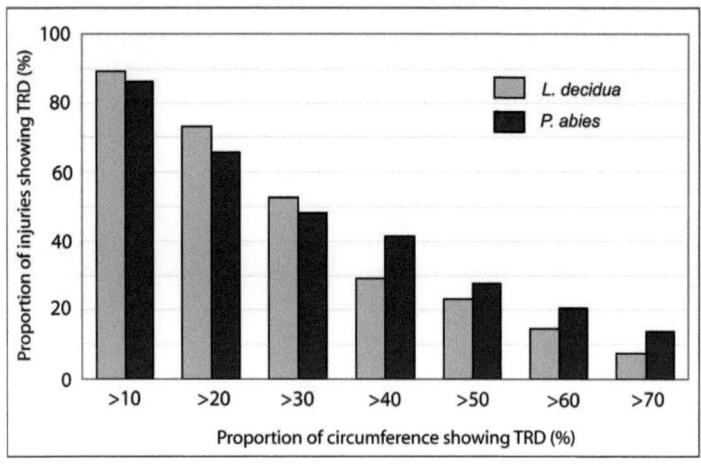

Figure 4. Radial spread of TRDs in that portion of the tree ring remaining vital in the first year following wounding.

Intra-annual shift of TRD

An intra-annual shift of TRDs with increasing tangential distance from the wound was observed in 73.2% of the *L. decidua* samples and 96.6% of the *P. abies* samples. There was a relationship between the tangential distance from the injury and the importance of the intra-annual shift of TRDs in *P. abies* (Figure 5), but no such dependence was observed in the *L. decidua* samples, indicating that distance is not the only factor that induces the intra-annual shift.

Persistence of TRD formation

The persistence of TRD formation in the years following the injuring events differed between the species. In the year of the event, both species reacted to the injury with the for-mation of tangential TRDs (Figure 6). The proportion of *L. decidua* showing TRDs was reduced by about 50% every year after injury. Thus, 1 year after wounding, 56.3% of the samples showed TRDs, 22.1% were showing TRDs 2 years after the impact, and 3 years later, TRDs were present in 10.5% of the samples. In contrast, TRD formation persisted much longer in *P. abies*, with ducts being present in 93.1% of all samples 1 year after the impact. Two years after the injuring event, TRDs were still produced in 69% of the samples and even 5 years after the disturbance, more than half of the *P. abies* (53.6%) samples had TRDs.

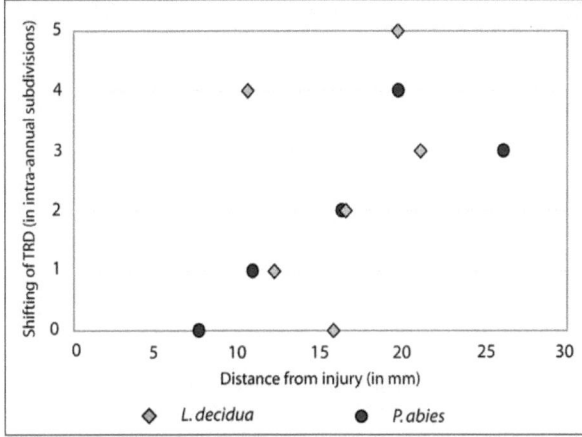

Figure 5. Relation between tangential distance from injury and radial migration of TRDs toward later portions of the tree ring (for the intra-annual subdivision, see Figure 2B).

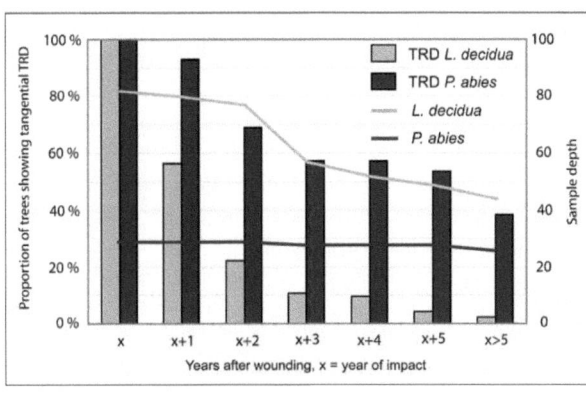

Figure 6. Persistence of TRD forma-tion in the years fol-lowing wounding in L. deci-dua (light gray) and P. abies (dark gray). The year of the impact is indicated with 'x'. The gray lines indi-cate the number of samples per spe-cies; the decline for L. deci-dua between x+2 and x+3 can be explained by the large number of injuries in 2004.

Discussion

We analyzed 111 rockfall injuries from 19 *L. decidua* Mill. and eight *P. abies* (L.) Karst. trees and assessed the anatomic changes in tree growth following wounding. The TRDs were always formed after wounding in the investigated samples. Most wounds (76.6%) were inflicted during the dormant season, resulting in TRD formation at the very beginning of the growth ring formed during the following growth period. These findings are in agreement with the studies by Bannan (1936), Fahn et al. (1979) and Bollschweiler et al. (2008).

It also appears from our data that TRD formation is the most prominent growth feature in *L. decidua* and *P. abies* after the rockfall impact. In the first ring that was formed after the impact, TRDs were observed in all samples. These findings corroborate those of Stoffel and Hitz (2008), who identified tangential TRDs in 94% of their rockfall samples. The tangential spread of TRDs is important, and ducts were found in 34% (*L. decidua*) and 36.4% (*P. abies*) of the circumference that was remaining vital after the impact (Figure 3). In contrast to the important tangential spread of TRDs, callus tissue was present on only 4.2% (*L. decidua*) and 3.6% (*P. abies*) of the functional tissue. These findings match the results of Bannan (1936), who mentioned that TRDs are present over 'considerable distances' in case of severe injuries. Our results are also in good agreement with the data obtained by Bollschweiler et al. (2008), who observed TRD formation in 19% of the vital circumference of *L. decidua* after debris-flow events. The differences in the tangential extent between the study by Bollschweiler et al. (2008) and this study are probably the result of the different nature of the wounding process. It is known that different geomorphic processes influence the tree response; for example, snow avalanche impacts do not cause the same reactions as rockfall injuries (Stoffel and Hitz 2008). The abrading action of debris flows can be compared with that of snow avalanches and they both induce a long-lasting and intense impact. In contrast, the energy transfer between a falling rock and a tree is transferred in only a fraction of a second, i.e., between 2 and 6×10^{-3} s (Rutz 1999) and energy is, moreover, concentrated at a single point on the tree's stem.

Our results also support the finding of Bannan (1936) that the intensity of a tree's response to a disturbance depends on the seasonality of the injuring event. We observed that the influence of seasonality was more obvious for callus tissue than for TRD formation. Callus tissue was formed much more frequently in trees during the growth period (42.3%) than during the dormant period (18.8%). In contrast, the difference was less obvious for TRD formation which showed a proportional increase from 32.7% to 40.7%.

The importance of TRD formation depends on different variables including tree age (Thomson and Sifton 1925, Bannan 1936) and injury size (Fahn et al. 1979, Bollschweiler et al. 2008). We tested the dependence of TRD formation in trees injured by rockfall on tree age and injury width, along with other variables such as DBH at the time of the impact as well as the annual DBH increment rate. For all variables, both tree species reacted in a comparable way. As shown in Figure 7A, we confirmed that TRD formation depends on injury width. The larger the wound, the more tangential TRD can be expected. This result contrasts with that of Bollschweiler et al. (2008) who had found no significant correlation between injury width and tangential spread of TRD.

We believe that this discrepancy between studies is associated with the nature of the disturbance, because injury size is presumably not a reliable indicator of the degree of the disturbance. As can be seen from Figure 7B, TRD formation is also influenced by the age of the tree at the time of wounding. Younger trees form fewer TRDs than older trees. These findings are consistent with those of Thomson and Sifton (1925) and Bannan (1936). Similarly, DBH at the time of injury affects tangential TRD extension; bigger trees are stiffer and less flexible and therefore show a stronger reaction (Figure 7C). This result is not surprising, because tree age and DBH are not independent variables; older trees are normally bigger than younger trees. Finally, annual DBH increment of a tree before injury did not influence TRD formation (Figure 7D).

The intra-annual shift of TRDs with increasing tangential distance from the injury has been reported in earlier studies. Bollschweiler et al. (2008) had reported a shift in one-third of their *L. decidua* samples following debris-flow activity. Stoffel and Hitz (2008) had confirmed these findings on

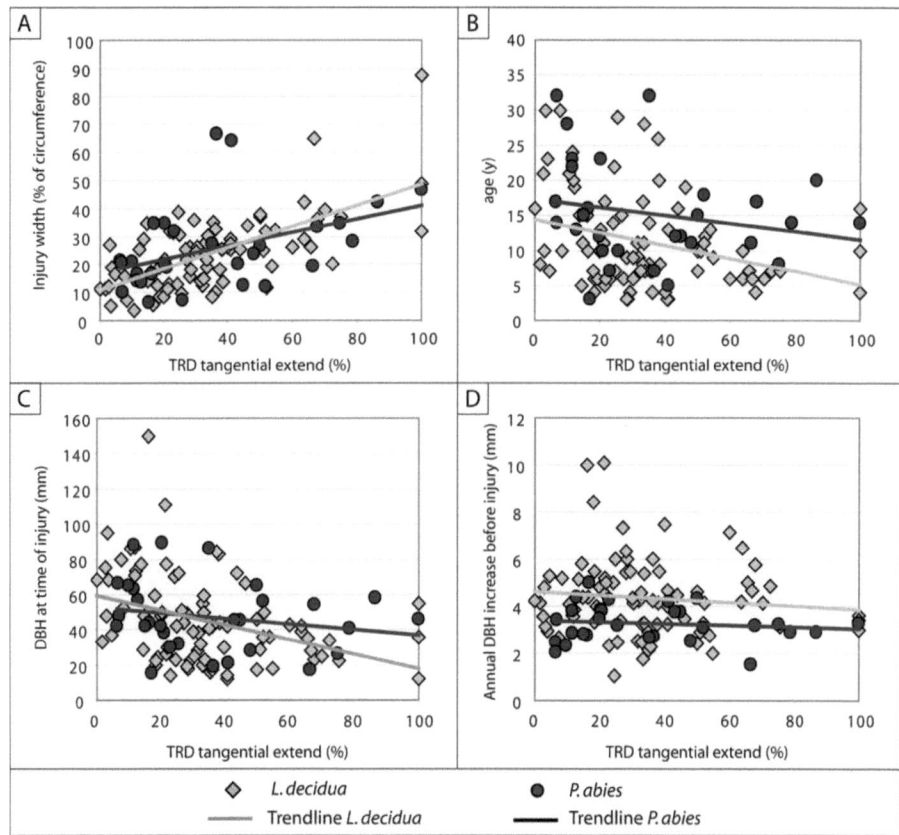

Figure 7. Dependence of TRD formation on (A) injury width; (B) tree age at the time of wounding; (C) DBH at the time of wounding; and (D) annual DBH increment rate before wounding, each with trendlines for L. decidua (gray) and P. abies (black).

L. decidua trees injured by snow avalanches and rockfalls, mentioning a shift in 36% of the snow avalanche samples and 31% of the rockfall samples. Such variations in the intra-annual position of TRDs are the result of a slowly propagating chemical signal in the tree after wounding. Krekling et al. (2004) had reported that this signal propagates about 2.5 cm day^{-1} in the axial direction in *P. abies*. In this study, 87.4% of the injuries showed an intra-annual shift with increasing distance from the wound, but the reason for this high proportion is unknown.

Tree reactions following rockfall impacts were similar in our study species. For example, there was less difference between the species in the tangential extension of callus tis-sue and TRDs, in the intra-seasonal shift of TRDs, or in the effects of tree age and injury width on TRD formation. There was a major difference between the species, however, in the persistence of TRD formation in the years following wounding. We found that the proportion of *L. decidua* trees showing TRDs was reduced by about 50% every additional year following wounding (Table 6). In contrast, the proportion of *P. abies* trees showing TRDs followed a more or less linear function. Five years after wounding, more than half of the *P. abies* samples still formed TRDs. A possible explanation for this diverging behavior of the species is a different resistance to mechanical impact. It seemed that injury to a *L. decidua* tree resulted in less stress than injury to a *P. abies* tree. A reason for this resilience might be that the relatively thick bark of *L. decidua* is able to absorb a certain amount of energy. Alternatively, because *L. decidua* is a pioneer plant that is able to grow under diffcult conditions, toleration of mechanical stress and accelerated healing might account for its high resistivity compared with *P. abies*.

The high proportion of trees having TRDs in the first ring that was formed after wounding is in agreement with the results of other studies (Table 6). No matter what geomorphic process caused the injury, at least 87.3% of the investigated samples presented TRDs as a response to wounding. However, TRD formation persisted for longer in *L. decidua* trees that were damaged by debris flows compared with trees injured by rockfall or snow avalanche activity.

Our observations on the tangential distribution of TRDs in rockfall trees show the potential of nondestructive sampling methods for

TRDs	*L. deciduas*[a]	*P. abies*[a]	*L. decidua*[b]	*L. decidua*[b]	*L. decidua*[c]
	Rockfall (%)	Rockfall (%)	Rockfall (%)	Avalanche (%)	Debris flow (%)
x	100	100	94.3	87.3	89.3
$x+1$	56.3	93.1	88.6	79.7	75
$x+2$	22.1	69	58.9	54.2	64.3
$x+3$	10.5	57.1	36.1	6.6	67.9
$x+4$	9.6	57.1	15.8	0.9	57.1
$x+5$	4.1	53.6	5.1	0	–
$x>5$	2.3	38.5	–	–	–

Table 6. Long-term appearance of tangential TRDs in L. decidua and P. abies following injury. Comparison of formation of tangential TRDs after wounding. Abbreviation: x, year of wounding.
Source: [a] Present study. [b] Stoffel and Hitz (2008). [c] Bollschweiler et al. (2008).

dendrogeomorphic studies. After detailed analyses of 111 rockfall wounds we can state that: (1) all samples produced TRDs next to the injuries and (2) the first appearance of TRDs was in the year of the injury or at the beginning of the new growing season when wounding occurred during the dormant season. We conclude that a reliable intra-annual dating of events is only possible if entire cross sections are analyzed. A careful sampling strategy based solely on increment cores with TRD as the marker allows rockfall injuries to be dated with a yearly precision.

In conclusion, our results clearly demonstrate the high potential of TRD analysis in dating past rockfall events. Nondestructive coring methods allow analysis of rockfall frequencies and can be used for the assessment of hazard and risks. This study also revealed a very high complexity of TRD formation after collision and identified some of the key factors that influence TRD formation. However, more research is needed to fully decode all aspects of TRD formation in trees.

Acknowledgments

The authors thank Lautaro Correa for his assistance in the field and Oliver Hitz for his helpful comments on wood anatomy. The authors express their appreciation to the local administration and the forest warden, Urs Andenmatten, who allowed them to work in the forest stand and gave permission to fell the experimental trees.

References

Bannan, M.W. 1936. Vertical resin ducts in the secondary wood of the Abietineae. New Phytol. 35:11–46.

Berger, F., C. Quetel and L.K.A. Dorren. 2002. Forest: a natural protection mean against rockfall but with which effciency? The objectives and methodology of the ROCKFOR project. Intrapraevent 2002 Band 2:815–826.

Bollschweiler, M. 2007. Spatial and temporal occurrence of past debris flows in the Valais Alps – results from tree-ring analysis. GeoFocus 20:182.

Bollschweiler, M., M. Stoffel, D.M. Schneuwly and K. Bourqui. 2008. Traumatic resin ducts in *Larix decidua* stems impacted by debris flows. Tree Physiol. 28:255–263.

Braam, R.R., E.E.J. Weiss and P.A. Burrough. 1987. Spatial and temporal analysis of mass movement using dendrochronology. CATENA 14:573–584.

Clark, G. 1981. Staining procedures. Williams and Wilkins, Baltimore and London, 512 p.

Fahn, A., E. Werker and P. Ben-Tzur. 1979. Seasonal effects of wounding and growth substances on development of traumatic resin ducts in Cedrus libani. New Phytol. 82:537–544.

Franceschi, V.R., T. Krekling and E. Christiansen. 2002. Application of methyl jasmonate on *Picea abies* (Pinaceae) stems induces defense-related responses in phloem and xylem. Am. J. Bot. 89:578–586.

Friedman, J.M., K.R. Vincent and P.B. Shafroth. 2005. Dating floodplain sediments using tree-ring response to burial. Earth Surf. Process. Landforms 30:1077–1091.

Hudgins, J.W., E. Christiansen and V.R. Franceschi. 2004. Induction of anatomical based defense responses in stems of diverse conifers by methyl jasmonate: a phylogenetic perspec-tive. Tree Physiol. 24:251–264.

Krekling, T., V.R. Franceschi, P. Krokene and H. Solheim. 2004. Differential anatomical response of Norway spruce stem tissues to sterile and fungus infected inoculations. Trees 18:1–9.

LePage, H. and Y. Bégin. 1996. Tree-ring dating of extreme water level events at Lake Bienville, Subarctic Québec, Canada. Arct. Alp. Res. 28:77–84.

Luchi, N., R. Ma, P. Capretti and P. Bonello. 2005. Systemic induction of traumatic resin ducts and resin flow in Austrian pine by wounding and inoculation with *Sphaeropsis sapinea* and *Diplodia scrobiculata*. Planta 221:75–84.

Mattheck, G.C. 1996. Trees – the mechanical design. Springer-Verlag, Berlin, 121 p.

Moore, K.W. 1978. Barrier-zone formation in wounded stems of sweetgum. Can. J. For. Res. 8:389–397.

Müller, H.N. 1980. Jahrringwachstum und Klimafaktoren: Beziehungen zwischen Jahrringwachstum von Nadelbaumarten und Klimafaktoren an verschiedenen Standorten im Gebiet des Simplonpasses (Wallis, Schweiz). Veröffentlichungen Forstliche Bundesversuchsanstalt Wien 25, Agrarverlag, Wien, 81 p.

Nagy, N.E., V.R. Franceschi, H. Solheim, T. Krekling and E. Christiansen. 2000. Wound-induced traumatic resin duct formation in stems of Norway spruce (Pinaceae): anatomy and cytochemical traits. Am. J. Bot. 87:313–320.

Rutz, J. 1999. Block-Anprall auf Stahlbetonwände aus Steinschlägen, Lawinen, Murgängen und Überschwemmungen Gebäudeversicherungsanstalt des Kantons. St. Gallen, Sankt Gallen, Switzerland (in German).

Schneuwly, D.M. and M. Stoffel. 2008a. Tree-ring based reconstruction of the seasonal timing, major events and origin of rockfall on a case-study slope in the Swiss Alps. Nat. Hazards Earth Syst. Sci. 8:203–211.

Schneuwly, D.M. and M. Stoffel. 2008b. Spatial analysis of rockfall activity, bounce heights and geomorphic changes over the last 50 years – a case study using dendrogeomor-phology. Geomorphology 102:522–531.

Schweingruber, F.H. 1978. Mikroskopische Holzanatomie. Flück, Teufen, Switzerland, 226 p.

Schweingruber, F.H. 1996. Tree Rings and Environment. Dendroecology, Paul Haupt, Bern, Switzerland, 188 p.

Schweingruber, F.H. 2001. Dendroökologische Holzanatomie. Paul Haupt, Bern, Switzerland, 472 p (in German).

Shigo, A.L. 1984. Compartmentalization: a conceptual frame-work for understanding how trees grow and defend themselves. Annu. Rev. Phytopathol. 22:189–214.

SMI (Swiss Meteorological Institute). 2008. Annals of the Swiss Meteorological Institute, daily precipitation sums and summarized rainfall, 1987–2007, Zurich.

Stoffel, M. 2008. Dating past geomorphic processes with tangential rows of traumatic resin ducts. Dendrochronologia 26:53–60.

Stoffel, M. and O.M. Hitz. 2008. Rockfall and snow avalanche impacts leave different anatomical signatures in tree rings of juvenile *Larix decidua*. Tree Physiol. 28:1713–1720.

Stoffel, M. and S. Perret. 2006. Reconstructing past rockfall activity with tree rings: some methodological considerations. Dendrochronologia 24:1–15.

Stoffel, M., I. Lièvre, M. Monbaron and S. Perret. 2005. Seasonal timing of rockfall activity on a forested slope at Täschgufer (Valais, Swiss Alps) – a dendrochronological approach. Zeitschrift fu"r Geomorphologie 49(1):89–106.

Strunk, H. 1997. Dating of geomorphological processes using dendrogeomorphological methods. CATENA 31:137–151.

Thomson, R.B. and H.B. Sifton. 1925. Resin canals in the Canadian spruce (*Picea canadensis* (Mill.) B.S.P.). Philos. Trans. R. Soc. Lond. B 214:63–111.

Timell, T.E. 1986. Compression wood in gymnosperms. Springer-Verlag, Berlin, 2150 p.

published in Tree Physiology 2009, Vol. 29, 1247–1257

Three-dimensional analysis of the anatomical growth response of European conifers to mechanical disturbance

Dominique M. Schneuwly [1,2], Markus Stoffel [1,3,4], Luuk K.A. Dorren [5], and Frédéric Berger [6]

[1] *Department of Geosciences, Geography, University of Fribourg, Chemin du Muse´e 4, CH-1700 Fribourg, Switzerland*

[2] *Corresponding author (dominique.schneuwly@unifr.ch)*

[3] *Laboratory of Dendrogeomorphology, Institute of Geological Sciences, University of Berne, Baltzerstrasse 1+3, CH-3012 Berne, Switzerland*

[4] *Environmental Sciences, University of Geneva, Chemin de Drize 7, CH-1227 Carouge-Geneva, Switzerland*

[5] *Federal Office for the Environment, Hazard Prevention Division, CH-3003 Bern, Switzerland*

[6] *Cemagref Grenoble, 2 Rue de la Papeterie, B.P. 76, F-38402 Saint Martin d'Hères Cedex, France*

Summary

Studies on tree reaction after wounding were so far based on artificial wounding or chemical treatment. For the first time, type, spread and intensity of anatomical responses were analyzed and quantified in naturally disturbed *Larix decidua* Mill., *Picea abies* (L.) Karst. and *Abies alba* Mill. trees. The consequences of rockfall impacts on increment growth were assessed at the height of the wounds, as well as above and below the injuries. A total of 16 trees were selected on rockfall slopes, and growth responses following 54 wounding events were analyzed on 820 cross-sections. Anatomical analysis focused on the occurrence of tangential rows of traumatic resin ducts (TRD) and on the formation of reaction wood. Following mechanical disturbance, TRD produc-tion was observed in 100% of *L. decidua* and *P. abies* wounds. The radial extension of TRD was largest at wound height, and they occurred more commonly above, rather than below, the wounds. For all species, an intra-annual radial shift of TRD was observed with increasing axial distance from wounds. Reaction wood was formed in 87.5% of *A. alba* following wounding, but such cases occurred only in 7.7% of *L. decidua*. The results demonstrate that anatomical growth responses following natural mechanical disturbance differ significantly from the reactions induced by artificial stimuli or by decapitation. While the types of reactions remain comparable between the species, their intensity, spread and persistence disagree considerably. We also illustrate that the external appearance of wounds does not reflect an internal response intensity. This study reveals that disturbance induced under natural conditions triggers more intense and more widespread anatomical responses than that induced under artificial stimuli, and that experimental laboratory tests considerably underestimate tree response.

Keywords

Abies alba, *Larix decidua*, *Picea abies*, reaction wood rockfall, tangential rows of traumatic resin ducts, wood anatomy.

Introduction

In Alpine forest ecosystems, intrinsic factors, such as climatic conditions, concurrence and age structure, constantly influence tree growth. A large variety of disturbing agents and their effects on tree growth and wood anatomy have been investigated in detail. Viveros-Viverosa et al. (2009) studied the effect of altitude on trees. The influence of soil property on tree growth was studied by Oberhuber et al. (1997), and the influence of temperature and precipitation was studied by Wimmer and Grabner (1997, 2000) and Esper et al. (2008). Rigling et al. (2003) analyzed the benefits of irrigation, whereas the changes in radial growth following water-level fluctuations were investigated by Polacek et al. (2006). Oberhuber et al. (1998), Rigling et al. (2002) and Weber et al. (2007) investigated the impact of drought. Tree disturbance due to pest infestation is yet another field of dendroecology that has received attention over the last few years (Weber 1997, Baier et al. 2002). The effect of soil erosion on tree growth assessed by McAuliffe et al. (2006), Hitz et al. (2008) and Spatz and Bruechert (2000) described the impact of wind factors.

In Alpine environments, another major disturbance influencing tree growth is mechanical stress. Macroscopic tree responses to mechanical impact are wounding, tilting and breaking of the stem (Schweingruber 1996). Macroscopic tree response following mechanical disturbance was analyzed in the past using resistance tests – i.e., artificial stem bending, (Moore 2000, Peltola et al. 2000, Stokes et al. 2005, Lundström et al. 2007a, 2008) so as to develop energy dissipation models or to obtain maximum stem breaking values. More recently, mechanical full-scale impact experiments were conducted in the field by Dorren and Berger (2005), Dorren et al. (2006) and Lundström et al. (2009) to determine the maximum energy dissipation values of different trees.

On a microscopic scale, trees respond to mechanical disturbance with the formation of callus tissue (Schweingruber 2001) and eccentric growth or with the formation of reaction wood (Timell 1986, Duncker and Spiecker 2008). In addition, different conifer species react to mechanical stress with the formation of tangential rows of traumatic resin ducts (TRD; Bannan 1936, Stoffel 2008). The TRD formation operates as a defence mechanism and assists compartmentalization of the wood after tree damage (Thomson and Sifton 1925, Shigo 1984, Hudgins et al. 2004). Resin ducts appear in tangential rows close to the wound, where the cambium had been partially destroyed due to mechanical stress. The arrangement of TRD is narrowest close to the wound and becomes more dispersed with increasing distance from it (Bannan 1936, Schneuwly et al. 2008). To date, only a very limited number of quantitative studies exist, which focus on the tangential or axial spread of TRD following mechanical stress or artificial decapitation on secondary xylem formation (e.g., Moore 1978, Nagy et al. 2000, Lev-Yadun 2002, Bollschweiler et al. 2008, Schneuwly et al. 2008, Stoffel and Hitz 2008). Another barely studied characteristic of TRD is the intra-annual shift of the ducts within a growth ring with increasing distance from the wound. Bannan (1936) described this phenomenon of delayed onset of TRD differentiation in an axial direction. Radial migration of TRD with increasing distance from the injury was recently confirmed by Bollschweiler et al. (2008), Schneuwly and Stoffel (2008a) and Stoffel and Hitz (2008).

Research has so far focused on the assessment of physical parameters in real-size experiments or on the anatomical growth features following artificial wounding of juvenile plants. In contrast, knowledge of the anatomical response following mechanical disturbance in adult trees is still largely needed, and the type, temporal appearance and spatial extent of tree response in the secondary xylem after natural wounding have not yet been addressed in detail.

This study, therefore, focuses on the types and the intensities of anatomical growth responses of European larch (*Larix decidua* Mill.), Norway spruce [*Picea abies* (L.) Karst.] and Silver fir (*Abies alba* Mill.) following mechanical disturbance induced by the impact of blocks falling from rock faces. The specific aims of this study were to (1) quantify the mean radial and axial extent of rows of tangential resin ducts; (2) investigate the spatial probabilities of the occurrence of TRD; (3) quantify the intra-annual shift of TRD in the axial direction; and (4) analyze the spatial distribution and the intensity of reaction wood.

Materials and methods

Study sites

The response of conifers to mechanical disturbance was analyzed in trees affected by natural rockfall activity. While each of the species was sampled on a different rockfall slope in the Swiss and French Alps, the determining conditions, such as slope angle and boulder size, did not differ considerably

Species	L. decidua	P. abies	A. alba
Location	Täsch	Saas Balen	Vaujany
Coordinates	46°04' N and 7°47' E	46°09' N and 7°55' E	45°12' N and 6°03' E
Slope angle	35°	36°	38°
Altitude	1800–1850 m	1550–1650 m	1250–1350 m
Boulder size	About 1 m^3	Maximum 1 m^3	About 1 m^3
Reference	Stoffel et al. (2005a)	Schneuwly and Stoffel (2008b)	Dorren et al. 2006

Table 1. Characteristics of the three different study areas of this study.

between the sites (Table 1). The altitude difference between the three study sites was a given precondition for this study.

Fieldwork and sample preparation

In total, 16 trees with multiple rockfall injuries were selected for the analysis: six *L. decidua*, six *P. abies* and four *A. alba*. For the sampled trees, we recorded tree species, tree height and tree diameter at breast height. We noted the maximum extension, width, height and orientation of each injury before it was photographed. The selected trees were then felled and cut into 820 stem sections of 0.1^{-m} length in the laboratory; they were then air dried and sanded up to grit 400.

While the mean tree diameters of *L. decidua* and *P. abies* were similar in size (18.7 and 23.7 cm, respectively), those of *A. alba* trees were much larger (mean diameter 43.1 cm). The average tree ages were 30.4 (*L. decidua*), 20.0 (*P. abies*) and 43.0 (*A. alba*) years (Table 2).

	Number of trees	Diameter (cm)		Age (years)	
		Mean	SD	Mean	SD
Total	16	24.2	39.3	28.4	15.8
L. decidua	6	18.7	24.9	30.4	17.1
P. abies	6	23.7	23	20	9.1
A. alba	4	43.1	53.5	43	12.9

Table 2. Number of sampled trees, tree diameters and tree age at the time of wounding.

Laboratory analysis

In the laboratory, wood-anatomical features relating to mechanical disturbance were identified using dendrogeomorphic methods described by Stoffel and Bollschweiler (2008) and Stoffel (2008). As a first step, the section with the maximum lateral extent of the injury was analyzed, followed by the sections above and below the exposed part of the wound insofar as any anatomical growth response could be detected. The width (in degrees) and the intra-annual position of the injury were assessed by noting the earliest evidence of the growth response, such as any change in the structure of tracheids (Stoffel and Hitz 2008), the formation of callus tissue or the appearance of TRD. The TRD were taken into account if they were present (i) in an extremely compact arrangement and (ii) forming continuous rows (Stoffel et al. 2005a). For the purpose of assessment of the intra-annual timing of wounding, responses in growth were classified following Stoffel et al. (2005b, Figure 1A) into: the first-formed cell layer (resulting from an injury during the dormant season, D), early, middle and late earlywood (EE, ME and LE, respectively, three equal-width sub-units) and early and late latewood (EL and LL, respectively, two equal-width sub-units).

The tangential extension of TRD at the maximum extent of wounds was calculated by measuring the distance between the wound's boundary and the farthest point where the TRD still occurred. Following Bollschweiler et al. (2008), the radial distance of TRD formation was given as a percentage of the ring's total circumference, excluding that portion where the cambium had been destroyed (Figure 1B). Above and below the injury, the radial spread of TRD was analyzed as well, but the values were divided by 2 to obtain data for both sides of the wound. In the next analytical step, the intra-annual season of the earliest tree response was noted, and the axial shift of TRD with increasing vertical distance from the wound was assessed. For this analysis, we only considered injuries showing first growth responses in the dormant season (D) and tracked the earliest appearance of growth responses in all sections above and below the wound. So as to provide representative results, at least 10 responses had to be pres-ent at each vertical distance from the wound to be considered.

The analysis of TRD formation concentrated particularly on the first year of appearance, but its presence in subsequent years was also noted. In the last analytical step, the occurrence of reaction wood was investigated, and its intensity was divided into three classes, with class 1 representing weak (weak appearance, little spread and few years), class 2 medium and class 3 intense (strong appearance, wide spread and several years) formation of reaction wood (Figure 2).

As some of the rockfall impacts were located close to ground level, it was not always possible to track growth responses to the level where they would have disappeared if they had occurred at a higher point on the stem. Similarly, anatomical growth responses

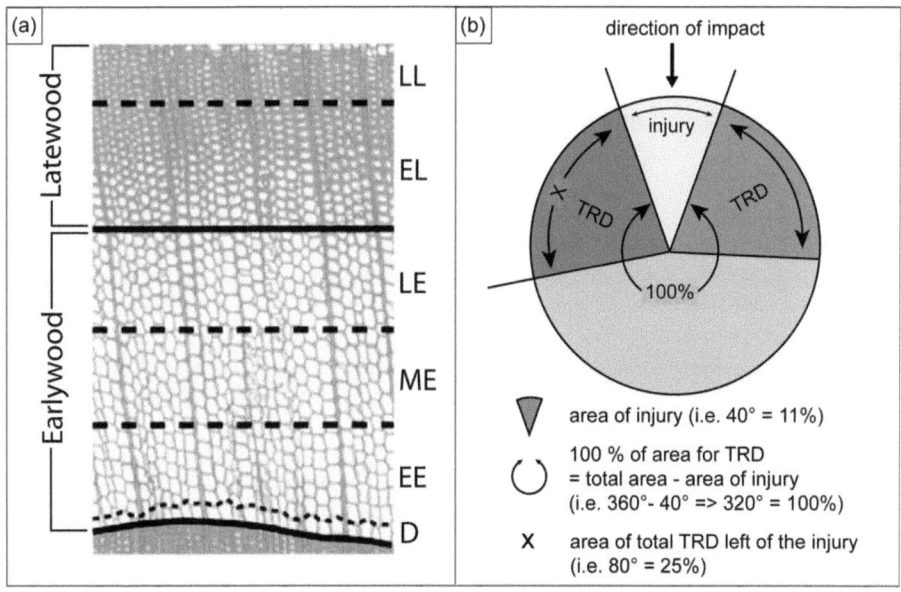

Figure 1. (A) Intra-annual subdivisions of a growth ring into dormancy (D), early earlywood (EE), middle earlywood (ME), late earlywood (LE), early latewood (EL) and late latewood (LL). (B) Schematic illustration of the tangential TRD analysis at the height of injury.

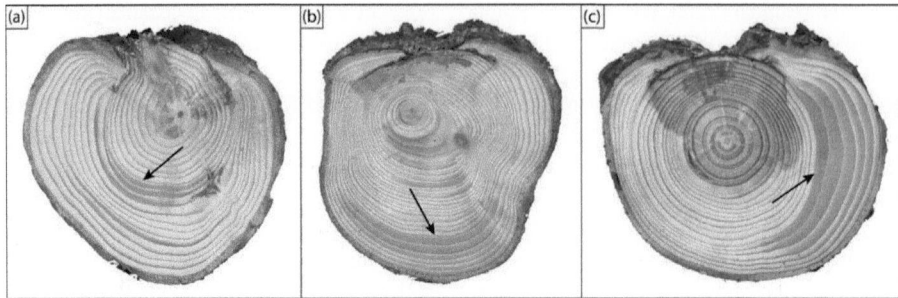

Figure 2. Illustration of the reaction wood intensity classes represents: (A) class 1 (weak), (B) class 2 (medium) and (C) class 3 (intense). The black arrows point at the first year of reaction wood formation. As can be seen in (C), the reaction wood does not necessarily appear opposite the direction of the injury.

were recorded only if they could be clearly attributed to a specific wound; in the case of proximity to another wound, ambiguous tree sections were not analyzed, and 'absent' sections were systematically removed from our calculations.

implying that the highest rockfall activity does not occur during the vegetation period. In *L. decidua*, 88.5% of all injuries occurred outside the vegetation period, whereas scarring in dormancy was found to be 60% in *P. abies* and 100% in *A. alba*.

Results

Characteristics of rockfall wounds

The impacts of rockfall resulted in 54 wounds on the investigated trees, meaning that most of the trees showed multi-ple wounds. Most of them were identified in *L. decidua* with 26, followed by *P. abies* with 20 and *A. alba* with 8 wounds. The mean wound height was 86.5 cm and the standard deviation (SD) was 88.3 cm (Table 3). The seasonality of wounding reveals a strong peak in D (79.6%),

Radial and axial arrangements of TRD in the year of injury

In the uninjured tissues adjacent to the wounds, the radial extent of TRD strongly depends on the species (Table 4). At the height of the wound, TRD were observed in all *L. decidua* and *P. abies* trees. In the case of *A. alba*, however, TRD were not present in two of eight injuries. The spread of the mean TRD proportion of the tree circumference where the cambium has not been destroyed again varied among the species.

	Number of wounds	Wound height (cm)		Intra-annual season of wounding					
		Mean	SD	D	EE	ME	LE	EI	LL
Total (%)	54	86.5	88.3	79.6	14.8	1.9	1.9	1.9	1.9
L. decidua (n)	26	91.2	92.8	23	1	1	0	1	0
P. abies (n)	20	93	91.6	12	7	0	1	0	0
A. alba (n)	8	55	64.6	8	0	0	0	0	0

Table 3. Number of wounds, wound heights and intra-seasonal distribution of wounding.

	Samples	Minimum	Maximum	Mean	SD
Radial arrangement (%)					
L. decidua	26	3.8	59	20.5	18.2
P. abies	20	4.5	100	27	22.8
A. alba	8	0	4.2	3.7	3.3
Axial arrangement (cm)					
L. decidua					
Above	25	0	60	22	13.5
Below	22	0	−50	−12.3	11.1
P. abies					
Above	20	10	130	38	31.2
Below	11	10	−60	−35.5	15.7
A. alba					
Above	8	0	20	10	9.3
Below	4	0	−10	−2.5	5

Table 4. Radial (%) and axial (cm) arrangements of TRD in L. decidua, P. abies and A. alba trees in the tree ring following mechanical disturbance.

While TRD in *L. decidua* and *P. abies* can be found on 20.5% and on 27%, respectively, its occurrence was only very localized in *A. alba* with 3.7% of the circumference. The analysis of the axial extension reveals tendencies similar to those observed for the radial arrangement, with TRD extensions above and below, respectively, the injury of +22 and −12.3 cm for *L. decidua*, +38 and −35.5 cm for *P. abies* and +10 and −2.5 cm for *A. alba*. The maximum axial extension for L decidua accounts for +60 and −50 cm. The TRD in *P. abies* was identified to be maximal up to +130 cm/−60 cm away from the wound. The smallest values are again noted for *A. alba*, with maximum vertical distances at +20 cm/−10 cm.

The arrangement of TRD in the radial direction is illustrated in Figure 3 and is given as the mean radial proportion affected at each 10-cm interval from the central height of the injury. The lateral and axial spread of TRD is largest in *P. abies*, followed by *L. decidua* and *A. alba*, and the highest values are observed just above the injury. The results also indicate that more TRD are normally formed above, rather than below, the injury after mechanical disturbance. The comparatively low values for the radial extent of TRD at the height of the injury are given because the proportion of the circumference occupied by the wound itself cannot produce TRD, and therefore does not appear in Figure 3.

Probability for TRD being formed in the year of injury

Figure 4 illustrates the probability of TRD being formed in the stem following mechanical disturbance. The highest probability for the occurrence of widespread TRD formation is observed in *P. abies* (Figure 4A). In the immediate lateral and upper neighborhood of the injury, the probability of TRD being formed exceeds 90% in *P. abies*. It also appears that the probability of TRD formation is generally higher above rather than below the injury in *P. abies*, but the differences in occurrence probabilities become much more comparable with increasing distance from the wound. Similar probability patterns can be observed for *L. decidua*, where the chances of TRD formation are again > 90% in the lateral and axial proximity of the wound (Figure 4B). When compared to the TRD pattern of *P. abies*, *L. decidua* shows

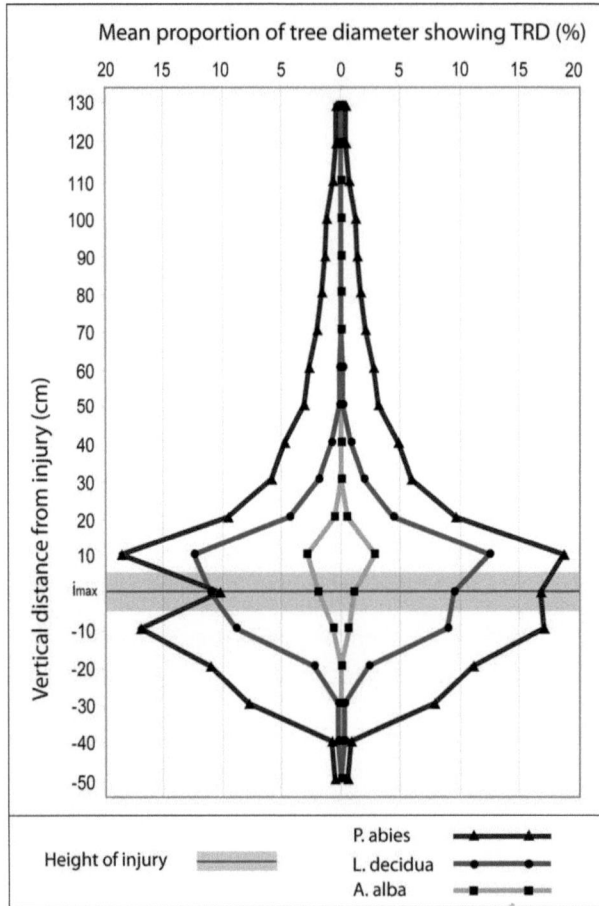

Figure 3. Mean spatial arrangement of TRD in the radial and axial directions for L. decidua, P. abies and A. alba following mechanical disturbance.

more rapid decreasing rates with increasing distance from the wound in both the radial and axial directions. Based on our data, it also appears that TRD in *L. decidua* is, in general, more commonly formed above, rather than below, the wound.

The probability of TRD being formed following mechan-ical disturbance is considerably smaller in *A. alba* than in the other two species (Figure 4C). The probability of find-ing TRD is again highest in the tissues adjacent to the wound at around 60–75%. At the height where the wounds exhibit their maximum radial extension, the probability of identifying TRD in 5% of the circumference remaining vital after disturbance already falls below 30%. Again, the probability of TRD formation is higher above rather than below the injury, and the slight radial decrease at the height of the injury is because the injury itself does not appear in the illustration.

Long-term production of TRD

Data on the long-term production of TRD are illustrated in Table 5 for the height of the injury as well as for the cross-sections just above and below the wound. At the height

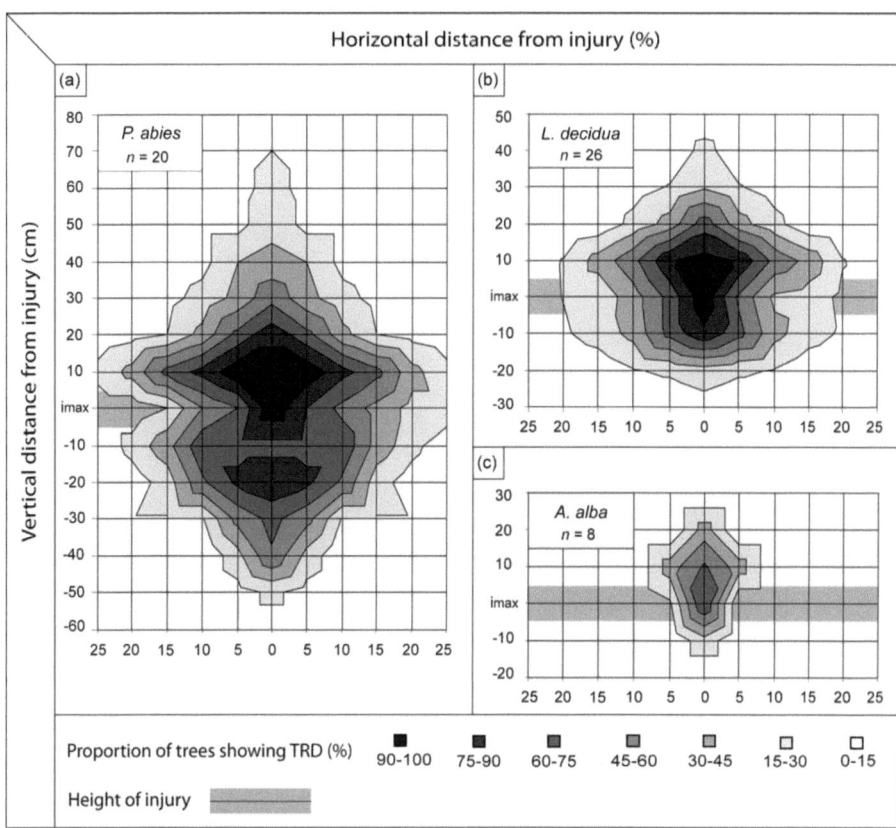

Figure 4. Probability of TRD being formed following mechanical disturbance in L. decidua, P. abies and A. alba. Values are given for the tree ring formed after mechanical disturbance.

of the injury, 69% of the natural disturbance events induced in *L. decidua* resulted in TRD formation for more than one growth period. In *P. abies*, 95% of the samples showed TRD for over > 1 year, whereas in *A. alba* 75% of the tissues adjacent to the wounds revealed TRD in the increment rings that were formed in the years after mechanical disturbance. The proportion of trees forming TRD above and below the wound for more than one season was, in contrast, considerably smaller for all species. Continuous TRD formation is most important for *P. abies*, where TRD is being produced for over 3.5 years after the year of impact at the height of the injury.

In *L. decidua* and *A. alba*, continuous TRD formation persists over 1.5 growth rings that are formed after the year of injury.

Intra-annual radial shift of TRD in axial direction

The intra-annual radial shift of TRD with increasing axial distance from the wound was investigated for all injuries induced before the onset of the local growing period in dormancy (D). As can be seen from Figure 5, an axial shift of TRD proves to be common for all species.

	Samples	TRD + y (N)	TRD + y (%)	Minimum (y)	Maximum (y)	Mean (y)	SD
L. decidua							
Above	25	12	48	0	11	1.1	2.3
Height injury	26	18	69.3	0	5	1.5	1.4
Below	23	9	39.1	0	2	0.6	0.8
P. abies*							
Above	20	16	80	0	9	2.3	2.2
Height injury	20	19	95	0	11	3.5	2.3
Below	14	9	64.3	0	11	2.9	3.1
A. alba*							
Above	8	2	25	0	1	0.3	0.5
Height injury	8	6	75	0	4	1.5	1.3
Below	4	1	25	0	1	0.3	0.5

Table 5. Long-term persistence of TRD in L. decidua, P. abies and A. alba following mechanical disturbance
y = Additional years of TRD formation, absolute (N) and proportion values (%) are given.
*Minimum values as some of the trees were observed before TRD formation ceased.

Although wounding occurred outside the vegetation period, and the earliest growth response was identified in the first tracheid cell layer of the new increment ring, we observed a systematic intra-annual radial shift to EE cell layers at only 10 cm above the injury for *L. decidua* and *A. alba*. In contrast, TRD formation in *P. abies* differs from those in *L. decidua* and *A. alba* for the cross-sections above the injury, where radial shifting is much more pronounced. At an axial distance of 10 cm above the wound, TRD are formed slightly in ME and approach LE cell layers at 30 cm.

Below the wound, the magnitude of intra-annual radial shifting is larger than that of above the injury, and similar values are

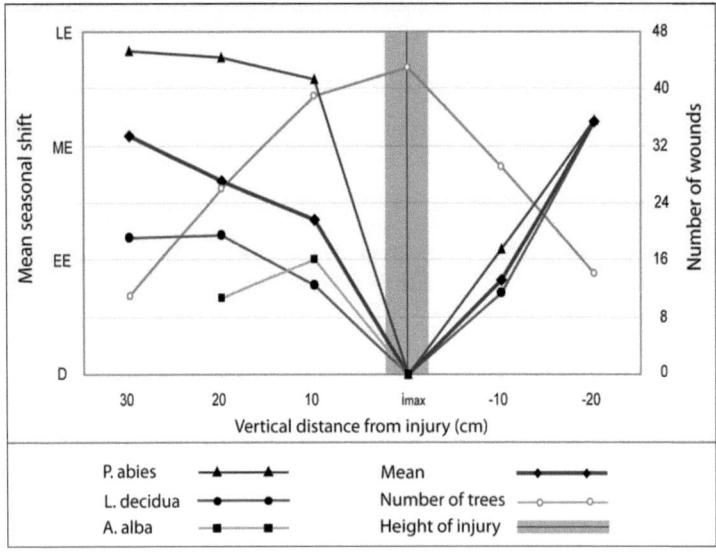

Figure 5. Axial intra-annual shift of TRD in L. decidua, P. abies and A. alba following mechanical disturbance.

observed between the species. In an axial distance of –10 cm below the wound, TRD appear in EE, whereas they are formed in ME and LE cell layers in an axial distance of –20 cm.

Arrangement and intensity of reaction wood following wounding

The axial position and the intensity of reaction wood were investigated, and the results are illustrated in Table 6. Interestingly, major differences exist between the species in appearance frequency and intensity. Simultaneously, reaction wood in all species can be found most commonly at the height of the injury. Reaction wood is also being formed in some of the segments above, but much less frequently below, the injuries. In *L. decidua*, reaction wood was pres-ent in only 7.7% of the samples at the height of the injuries, and the intensities were always small (class 1). In *P. abies*, in contrast, reaction wood seems to be more common and was observed in 50% of the samples at injury height. Reaction wood also exhibits a much more pronounced axial spread, with 50% of the samples above and with 42.9% of the samples below the wounds showing it. Moreover, the intensity in *P. abies* was more important than that in *L. decidua*, and regularly reached class 3.

Reaction wood seems to be a more frequent response to mechanical disturbance following rockfall impacts in *A. alba*, and was present in the vast majority (87.5%) of the samples at injury height. Above the wound, half of the segments still exhibited reaction wood, but none was observed below the injuries. The intensity of reaction wood largely varied in *A. alba*, and all classes were present on the samples.

Discussion

In this study, 54 injuries due to natural rockfall activity were analyzed in six *L. decidua*, eight *P. abies* and four *A. alba* trees

	Samples	Rw (N)	Rw (%)	N 1	N 2	N 3	Mean
L. decidua							
Above	26	1	3.9	1	0	0	0.04
Height injury	26	2	7.7	3	0	0	0.12
Below	23	0	0	0	0	0	0
P. abies							
Above	20	7	50	3	1	3	0.7
Height injury	20	7	50	1	3	3	0.8
Below	14	6	42.9	3	0	3	0.9
A. alba							
Above	8	4	50	3	1	0	0.6
Height injury	8	7	87.5	3	1	3	1.8
Below	4	0	0	0	0	0	0

Table 6. Axial position and intensity of reaction wood in L. decidua, P. abies and A. alba following mechanical disturbance.
Rw = Reaction wood, absolute (N) and proportion values (%) are given.
1–3 = Intensity of reaction wood (1 = weakest and 3 = strongest).

to identify and quantify anatomical growth response in the form of tangential rows of TRD and reaction wood following mechanical disturbance.

The results indicate that all *L. decidua* and *P. abies* responded to mechanical disturbance with the formation of TRD at the height of the injury. Similar to the results presented by Schneuwly et al. (2008), we observe TRD on one-fifth and one-fourth, respectively, of the circumference remaining vital after wood-penetrating impacts. In agreement with Schweingruber (2001), we also observe that formation in the radial direction in *A. alba* is much less pronounced (3.7%). The inhibited TRD formation of *A. alba* trees is not the consequence of a different mechanical behavior, but solely of the genetic makeup (Schweingruber 1996); *L. decidua* and *P. abies* also produce much more TRD above and below the wound.

It has also become evident that the mean axial extension of TRD following naturally induced mechanical disturbance (+22 and –12.3 cm for *L. decidua*, +38 and –35.5 cm for *P. abies*, +10 and –2.5 cm for *A. alba*) largely exceeds the values observed after artificial treatments, where the axial extension ranging from 5 cm (Franceschi et al. 2002) to 10 cm (Lev-Yadun 2002) or 12 cm (Luchi et al. 2005) has been reported. The considerably lower values of the studies cited above can be explained by the different nature of disturbance, as trees were artificially treated with hormones or decapitation in these studies and were not subjected to short-lived, but high-intensity, impacts. In contrast, our data seem to be comparable to the results reported by Bollschweiler et al. (2008), who observed TRD at vertical distances of +43 and –14 cm. Although similar to our trees, their samples were wounded by natural mechanical disturbance (debris flows).

The TRD formation in the axial direction was more important above, rather than below, the injury and therefore confirms the findings of Fahn et al. (1979) and Bollschweiler et al. (2008). We believe that the higher frequency of TRD above the wound results from an upward propagation of impact shock waves through the stem ('hula-hoop effect'; Dorren et al. 2006). Downward from the wound, the effects of shock waves may have been partially extenuated by the root system and the subsurface (Lundström et al. 2007b, 2007c). While this may explain the uneven axial distribution of TRD following natural mechanical disturbances, it may not elucidate the unequal arrangement of TRD after hormonal treatment, as reported by Fahn et al. (1979). As a result, it is also feasible that the signaling agents leading to TRD formation would more easily propagate in the upward direction.

It has also been demonstrated that TRD formation varies between the species and the nature of disturbance. Previous studies report on correlations between TRD formation and wound size (Fahn et al. 1979), wound size effects (area and length) on axial TRD extension (Bollschweiler et al. 2008) or on the influence of injury width on the radial extension of TRD (Schneuwly et al. 2008). We therefore tested the influence of injury size – categorized into three classes: smallest third, medium third and largest third – on TRD occurrence for *L. decidua* and *P. abies*. As can be seen in Figure 6A, the probability of TRD being formed in *L. decidua* is neither a function of size (Figure 6A) nor a function of maximum wound width (Figure 6B). Figure 6C and D illustrates that the findings for *P. abies* show a very similar response (the sample depth for *A. alba* did not allow for similar investigations).

The TRD in *P. abies* not only appeared more

extensively in the radial and axial directions and with higher probabilities, but they also persisted over a larger number of years after wounding. While our findings on *P. abies* and on *L. decidua* agree with the results of Schneuwly et al. (2008), it should also be mentioned that TRD formation was ongoing at the time of sampling in several *P. abies* and *A. alba* trees.

The intra-annual radial shift with increasing axial distance from the wound was detected in all species and confirmed the findings by Bannan (1936) and Bollschweiler et al. (2008). This delay in TRD formation might have resulted from the slow propagation of the signaling agent leading to resin production and resin duct formation. Krekling et al. (2004) state that, for *P. abies*, the signal propagates about 2.5 cm day^{-1} in the axial direction. Based on the phenological observations and growth data of *L. decidua* and *P. abies* from the wider study area (Müller 1980), we are led to believe that the formation of EE and ME tracheids lasts for around 30 days. As we had already observed a shift from the growth ring boundary to ME cell layers at 20–30 cm above and below the wound, we are led to assume that the propagation rates obtained after natural mechanical disturbance would be considerably smaller than those of the reactions after artificial treatment.

The formation of reaction wood does not represent a common growth reaction to

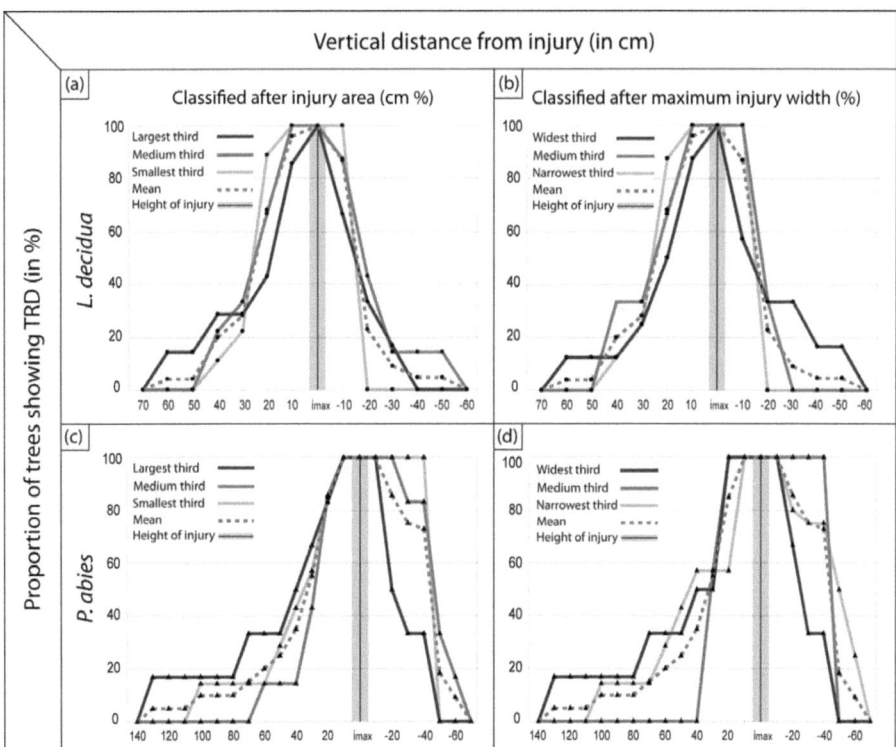

Figure 6. Spatial arrangement of TRD in L. decidua and P. abies following impact, classified after injury size and injury width.

mechanical disturbance in *L. decidua*, where it was present in only 7.7% of the sam-ples. Our findings agree with those of Stoffel et al. (2005a) and Schneuwly and Stoffel (2008a), who did not detect large amounts of the reaction either. In *P. abies*, in contrast, 50% of the samples showed reaction wood and at a higher intensity. Reaction wood was most commonly observed in *A. alba*, where it was detected in 87.5% of the samples. While soil characteristics and soil depths are very similar between the sites involved, we consider that the differences in the presence and in the intensity of reaction wood would result from the species-specific disparities in root system morphologies and related tree stability.

Conclusion

Growth response to mechanical disturbance resulting from natural rockfall was investigated in *L. decidua*, *P. abies* and *A. alba*. The results provide evidence that natural disturbance induces much stronger and widespread anatomical responses than any wounding produced through artificial stimuli. For the first time, the spatial arrangement, probability of occurrence and intra-annual radial shift of TRD could be quantified, as well as the position and the intensity of reaction wood formation following the impacts assessed. While this study revealed a multitude of new insights into anatomical growth responses of European conifers, it also raises new questions, namely the reasons for the preferential spread of resin ducts above injuries, and concerning the widespread absence of reaction wood in *L. decidua*.

Acknowledgments

The authors thank Dr. Michelle Bollschweiler for her assistance in the field and laboratory and for helpful comments on a previous version of this manuscript. They wish to acknowledge Dr. Oliver Hitz for his insights into wood anatomy and sample preparation. They also express their gratitude to the local foresters for their sup-port and sampling permission, and to Daniel Cuennet, Susanne Widmer, Nathalie Abbet, Pascal Tardif and Eric Mermin for their assistance in the field and the carpenter's shop. They finally thank Bill Harmer for proofreading this document.

References

Baier, P., E. Führer, T. Kirisits and S. Rosner. 2002. Defence reactions of Norway spruce against bark beetles and the associated fungus *Ceratocystis polonica* in secondary pure and mixed species stands. For. Ecol. Manag. 159:73–86.

Bannan, M.W. 1936. Vertical resin ducts in the secondary wood of the Abietineae. New Phytol. 35:11–46.

Bollschweiler, M., M. Stoffel, D.M. Schneuwly and K. Bourqui. 2008. Traumatic resin ducts in *Larix decidua* stems impacted by debris flows. Tree Physiol. 28:255–263.

Dorren, L.K.A. and F. Berger. 2005. Energy dissipation and stem breakage of trees at dynamic impacts. Tree Physiol. 26:63–71.

Dorren, L.K.A., F. Berger and U.S. Putters. 2006. Real-size experiments and 3-D simulation of rockfall on forested and non-forested slopes. Nat. Hazards Earth Syst. Sci. 6: 145–153.

Duncker, P. and H. Spiecker. 2008. Cross-sectional compression wood distribution and its relation to eccentric radial growth in *Picea abies* (L.) Karst. Dendrochronologia 26:195–202.

Esper, J., R. Niederer, P. Bebi and D. Frank. 2008. Climate signal age effects – evidence from young and old trees in the Swiss Engadin. For. Ecol. Manag. 255:3783–3789.

Fahn, A., E. Werker and P. Ben-Tzur. 1979. Seasonal effects of wounding and growth substances on development of traumatic resin ducts in *Cedrus libani*. New Phytol. 82:537–544.

Franceschi, V.R., T. Krekling and E. Christiansen. 2002. Application of methyl jasmonate on *Picea abies* (Pinaceae) stems induces defense-related responses in phloem and xylem. Am. J. Bot. 89:578–586.

Hitz, O.M., H. Gärtner, I. Heinrich and M. Monbaron. 2008. Wood anatomical changes in roots of European ash (*Fraxinus excelsior* L.) after exposure. Dendrochronologia 25:145–152.

Hudgins, J.W., E. Christiansen and V.R. Franceschi. 2004. Induction of anatomical based defense responses in stems of diverse conifers by methyl jasmonate: a phylogenetic perspective. Tree Physiol. 24:251–264.

Krekling, T., V.R. Franceschi, P. Krokene and H. Solheim. 2004. Differential anatomical response of Norway spruce stem tissues to sterile and fungus infected inoculations. Trees 18:1–9.

Lev-Yadun, S. 2002. The distance to which wound effects influence the structure of secondary xylem of decapitated Pinus pinea. J. Plant Growth Regul. 21:191–196.

Luchi, N., R. Ma, P. Capretti and P. Bonello. 2005. Systemic induction of traumatic resin ducts and resin flow in Austrian pine by wounding and inoculation with *Sphaeropsis sapinea* and *Diplodia scrobiculata*. Planta 221:75–84.

Lundström, T., U. Heiz, M. Stoffel and V. Stöckli. 2007a. Fresh-wood bending: linking the mechanical and growth properties of a Norway spruce stem. Tree Physiol. 27:1229–1241.

Lundström, T., M.J. Jonsson and M. Kalberer. 2007b. The root–soil system of Norway spruce subjected to turning moment: resistance as a function of rotation. Plant Soil 300:35–49.

Lundström, T., T. Jonas, V. Stöckli and W. Ammann. 2007c. Anchorage of mature conifers: resistive turning moment, root–soil plate geometry and root growth orientation. Tree Physiol. 27:1217–1227.

Lundström, T., M. Stoffel and V. Stöckli. 2008. Fresh-stem bending of fir and spruce. Tree Physiol. 28:355–366.

Lundström, T., M.J. Jonsson, A. Volkwein and M. Stoffel. 2009. Reactions and energy absorption of trees subject to rockfall: a detailed assessment using a new experimental method. Tree Physiol. 29:345–359.

McAuliffe, J.R., L.A. Scuderi and L.D. McFadden. 2006. Tree-ring record of hillslope erosion and valley floor dynamics: landscape responses to climate variation during the last 400 yr in the Colorado Plateau, northeastern Arizona. Global Planet. Change 50:184–201.

Moore, J.R. 2000. Differences in maximum bending moments of *Pinus radiata* trees grown on a range of soil types. For. Ecol. Manag. 135:63–71.

Moore, K.W. 1978. Barrier-zone formation in wounded stems of sweetgum. Can. J. For. Res. 8:389–397.

Müller, H.N. 1980. Jahrringwachstum und Klimafaktoren: Beziehungen zwischen Jahrringwachstum von Nadelbaumarten und Klimafaktoren an verschiedenen Standorten im Gebiet des Simplonpasses (Wallis, Schweiz). Veröoentlichungen Forstliche Bundesversuchsanstalt Wien 25. Agrarverlag, Wien, 81 p. (in German).

Nagy, N.E., V.R. Franceschi, H. Solheim, T. Krekling and E. Christiansen. 2000. Wound-induced traumatic resin duct formation in stems of Norway spruce (Pinaceae): anatomy and cytochemical traits. Am. J. Bot. 87:313–320.

Oberhuber, W., K. Pagitz and K. Nicolussi. 1997. Subalpine tree growth on serpentine soil: a dendroecological analysis. Plant Ecol. 130:213–221.

Oberhuber, W., M. Stumböck and W. Kofler. 1998. Climate-tree-growth relationships of Scots pine stands (*Pinus sylvestris* L.) exposed to soil dryness. Trees 13:19–27.

Peltola, H., S. Kellomäki, A. Hassinen and M. Granader. 2000. Mechanical stability of Scots pine, Norway spruce and birch: an analysis of tree-pulling experiments in Finland. For. Ecol. Manag. 135:143–153.

Polacek, D., W. Kofler and W. Oberhuber. 2006. Radial growth of *Pinus sylvestris* growing on alluvial terraces is sensitive to water-level fluctuations. New Phytol. 169:299–308.

Rigling, A., O.U. Bräker, G. Schneiter and F. Schweingruber. 2002. Intra-annual tree-ring parameters indicating differences in drought stress of *Pinus sylvestris* forests within the Erico-Pinion in the Valais (Switzerland). Plant Ecol. 163:105–121.

Rigling, A., H. Bruhlhart, O.U. Braker, T. Forster and F.H. Schweingruber. 2003. Effects of irrigation on diameter growth and vertical resin duct production in *Pinus sylvestris* L. on dry sites in the central Alps, Switzerland. For. Ecol. Manag. 175:285–296.

Schneuwly, D.M. and M. Stoffel. 2008a. Tree-ring based reconstruction of the seasonal timing, major events and origin of rockfall on a case-study slope in the Swiss Alps. Nat. Hazards Earth Syst. Sci. 8:203–211.

Schneuwly, D.M. and M. Stoffel. 2008b. Changes in spatio-temporal patterns of rockfall activity on a forested slope – a case study using dendrogeomorphology. Geomorphology 102:522–531.

Schneuwly, D.M., M. Stoffel and M. Bollschweiler. 2008. Formation and spread of callus tissue and tangential rows of resin ducts in *Larix decidua* and *Picea abies* following rockfall impacts. Tree Physiol. 29:281–289.

Schweingruber, F.H. 1996. Tree rings and environment. Dendroecology. Paul Haupt, Bern, Switzerland, 188 p.

Schweingruber, F.H. 2001. Dendroökologische Holzanatomie. Paul Haupt, Bern, Switzerland, 472 p. (in German).

Shigo, A.L. 1984. Compartmentalization: a conceptual frame-work for understanding how trees grow and defend them-selves. Annu. Rev. Phytopathol. 22:189–214.

Spatz, H.C. and F. Bruechert. 2000. Basic biomechanics of self-supporting plants: wind loads and gravitational loads on a Norway spruce tree. For. Ecol. Manag. 135:33–44.

Stoffel, M. 2008. Dating past geomorphic processes with tangential rows of traumatic resin ducts. Dendrochronologia 26:53–60.

Stoffel, M. and M. Bollschweiler. 2008. Tree-ring analysis in natural hazards research – an overview. Nat. Hazards Earth Syst. Sci. 8:187–202.

Stoffel, M. and O.M. Hitz. 2008. Rockfall and snow avalanche impacts leave different anatomical signatures in tree rings of juvenile *Larix decidua*. Tree Physiol. 28:1713–1720.

Stoffel, M., D. Schneuwly, M. Bollschweiler, I. Lie`vre, R. Delaloye, M. Myint and M. Monbaron. 2005a. Analyzing rockfall activity (1600–2002) in a protection forest – a case study using dendrogeomorphology. Geomorphology 68: 224–241.

Stoffel, M., I. Lièvre, M. Monbaron and S. Perret. 2005b. Seasonal timing of rockfall activity on a forested slope at Täschgufer (Valais, Swiss Alps) – a dendrochronological approach. Zeitschrift Geomorphologie 49:89–106.

Stokes, A., F. Salin, A. Kokutse, S. Berthier, H. Jeannin, S. Mochan, L. Dorren, N. Kokutse, M. Abd.Ghani and T. Fourcaud. 2005. Mechanical resistance of different tree species to rockfall in the French Alps. Plant Soil 278: 107–117.

Thomson, R.B. and H.B. Sifton. 1925. Resin canals in the Canadian spruce (*Picea canadensis* (Mill.) B.S.P.). Philos. Trans. R. Soc. Lond. B214:63–111.

Timell, T.E. 1986. Compression wood in gymnosperms. Springer, Berlin, 2150 p.

Viveros-Viverosa, H., C. Sàenz-Romerob, J.J Vargas-Hernàn-deza, J. López-Uptona, G. Ramírez-Valverdec and A. Santacruz-Varelad. 2009. Altitudinal genetic variation in *Pinus hartwegii* Lindl. I: height growth, shoot phenology, and frost damage in seedlings. For. Ecol. Manag. 257:836–842.

Weber, U.M. 1997. Dendroecological reconstruction and interpretation of larch budmoth (*Zeiraphera diniana*) outbreaks in two central Alpine valleys of Switzerland from 1470–1990. Trees 11:277–290.

Weber, P., H. Bugmann and A. Rigling. 2007. Radial growth responses to drought of *Pinus sylvestris* and *Quercus pubescens* in an inner-Alpine dry valley. J. Veg. Sci. 18:777–792.

Wimmer, R. and M. Grabner. 1997. Effects of climate on vertical resin duct density and radial growth of Norway spruce (*Picea abies* (L.) Karst.). Trees 11:271–276.

Wimmer, R. and M. Grabner. 2000. A comparison of tree-ring features in *Picea abies* as correlated with climate. IAWA J. 21:403–416.

♦♦♦♦♦

Chapter C

Applied research

published in Natural Hazards and Earth System Sciences 2008, Vol. 8, 203–211

Tree-ring based reconstruction of the seasonal timing, major events and origin of rockfall on a case-study slope in the Swiss Alps

Dominique M. Schneuwly and Markus Stoffel

Laboratory of Dendrogeomorphology, Department of Geosciences, University of Fribourg, Chemin du Musée 4, 1700 Fribourg, Switzerland

Abstract

Tree-ring analysis has been used to reconstruct 22 years of rockfall behavior on an active rockfall slope near Saas Balen (Swiss Alps). We analyzed 32 severely injured trees (*L. decidua*, *P. abies* and *P. cembra*) and investigated cross-sections of 154 wounds.

The intra-annual position of callus tissue and of tangential rows of traumatic resin ducts was determined in order to reconstruct the seasonality of past rockfall events. Results indicate strong intra- and inter-annual variations of rockfall activity, with a peak (76%) observed in the dormant season (early October – end of May). Within the growth season, rockfall regularly occurs between the end of May and mid July (21.4%), whereas events later in the season appear to be quite rare (2.6%). Findings suggest that rockfall activity at the study site is driven by annual thawing processes and the circulation of melt water in preexisting fissures. Data also indicate that 43% of all rockfall events occurred in 1995, when two major precipitation events are recorded in nearby meteorological stations. Finally, data on impact angles are in very good agreement with the geomorphic situation in the field.

1 Introduction

Rockfall is the free fall, bouncing or rolling of individual or a few rocks and boulders with volumes involved generally limited to a few cubic meters (Berger et al., 2002). Due to its hazard potential (Butler, 1983; Evans and Hungr, 1993), rockfall has become one of the most intensely studied geomorphic processes of the cliff zone in mountain areas. As a result, there exist a large number of studies analyzing rockfall mechanics, such as the movement (Ritchie, 1963; Erismann, 1986; Azzoni et al., 1995), the behavior during ground contact (Bozzolo and Pamini, 1986; Hungr and Evans, 1988) or runout distances (Kirkby and Statham, 1975; Statham and Francis, 1986; Okura et al., 2000). Furthermore, there are a large number of studies on rockfall modeling (Guzzetti et al., 2002; Dorren et al., 2006; Stoffel et al., 2006b) or on long-term accretion rates of rockfall (Luckman and Fiske, 1995; McCarroll et al., 1998). In addition, research has focused on possible triggers of rockfall such as tectonic folding (Coe and Harp, 2007), freeze-thaw cycles (Gardner, 1983; Matsuoka and Sakai, 1999; Matsuoka, 2006), changes in the rock-moisture level (Sass, 2005), the thawing of permafrost (Gruber et al., 2004), the rising of mean annual temperatures (Davies et al., 2001) or the occurrence of earthquakes (Harp and Wilson, 1995; Marzorati et al., 2002).

In contrast, only a few studies have analyzed the evolution of rockfall activity with time or determined its seasonal behavior: In the 1970s, a number of observation-based studies was realized in North America (Luckman, 1976; Douglas, 1980; Gardner, 1980). Gardner's (1980) observations on rockfall and rockslides in Alberta (Canada) were, for instance, kept on a weekly basis over a period of two years. He observed a continuous, yet small background activity throughout the year with two peaks in rockfall activity in February–March and November–December. More recently, Sass (2005) installed more than 60 rockfall barriers in the Bavarian Alps so as to quantify rockfall activity over a period of four years. He concludes that the distribution of rockfall is highly variable in time and that rocks and boulders are triggered by a combination of a multitude of parameters.

While these studies yielded very precise datasets on daily variations of rockfall, they do neither represent complete time series within the year nor did they cover continuous period over several years. As a consequence, they have to be considered too short to assess rockfall activity with its seasonal variations over time.

In contrast to other mass-movement processes (e.g., Shroder, 1980; Braam et al., 1987; Wiles et al., 1996; Solomina, 2002), tree-ring analyses have only occasionally been used to study past rockfall activity, mainly to determine sedimentation rates on scree slopes (Lafortune et al., 1997) or to reconstruct rates or the spatial occurrence of rockfall (Stoffel et al., 2005a; Perret et al., 2006).

In a similar way, dendrogeomorphology has rarely been applied to assess the seasonality of past mass-movement processes so far. Through the analysis of injuries and the adjacent callus tissue, Stoffel et al. (2005b) determined the seasonal timing of rockfall on a slope in the Valais Alps. Their results clearly indicate a peak in rockfall activity between October and May. More recently, the intra-annual position of tangential rows of traumatic resin ducts was used to separate previous snow avalanche from debris-flow events on a cone affected by both processes (Stoffel et al., 2006a) or to date past debris-flow activity with monthly precision (Stoffel et al., 2008).

It is therefore the aim of this study to close the gap between long-term but low seasonal and short-term but high seasonal resolution rockfall research by accomplishing a long-term study with a high seasonal resolution. Through the analysis of trees injured by rocks and boulders, we (i) reconstruct rockfall activity and possible event years; (ii) analyze the sea-sonal behavior of rockfall during the last 22 years so as to detect possible seasonal shifts of activity and (iii) compare the angles of the rockfall injuries as observed on the stem surface with the effective source area of falling rocks. Results were obtained from 123 cross-sections of 23 European larch (*Larix decidua* Mill.), eight Norway spruce (*Picea abies* (L.) Karst.) and one Swiss stone pine (*Pinus cembra* L.) trees growing in the transition zone of a rockfall slope in the Swiss Alps.

2 Study area

The study was conducted in a forest near Saas Balen (46° 09' 06" N, 7° 55' 27" E) in the Valais Alps (Switzerland; Fig. 1A). The investigated forest stand is called "Schilt" and is located between 1470 and 1610 m a.s.l. on the east-northeast facing slope descending from the Lammenhorn (3189 m a.s.l.). Rockfall frequently occurs on the slope, originating from disintegrated and glacially oversteepened cliffs at 1600–1900 m a.s.l (Figs. 1B and 2). In the departure zone, bedrock consists of micaceous schists belonging to Penninic crystalline layers, dipping SSE with an angle of 20° (Bearth, 1973, 1980). The transition and deposition zones are covered with Quaternary talus and morainic deposits. Archival data and local toponomy indicate the presence of rockfall in the region since at least the early 18th century, when rockfalls descended from the neighboring "Steinschlagwald" (= rockfall forest), destroying the old communal church (Ruppen et al., 1979). In contrast, other mass-movement processes such as debris flows or snow avalanches have neither ever been witnessed on the slope, nor is there geomorphic evidence for such processes.

On the study site, the volume of rocks resting on the talus slope does not normally exceed 1 m^3, but there are a few blocks deposited on

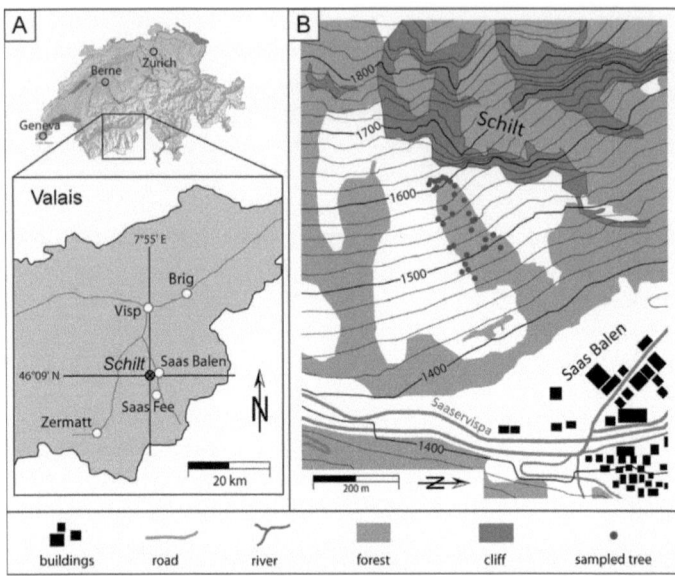

Fig. 1. (A). The study site is located in the Saas Valley near Saas Balen. (B). Sketch map of the study site and the position of the 32 trees sampled for analysis.

the valley bottom with volumes up to 50 m³, witnessing of major events in the past. The mean slope angle of the transition zone is 36° (Fig. 2C) with only little variation between the top and the bottom. The forested study site covers about 8000 m² and past rockfall deposits are covered with a centimetric soil layer. The zones outside the forest remain mostly free of vegetation and are covered with bare rocks and boulders. The stand at Schilt consists predominantly of *L. decidua* trees, accompanied by approximately 10% of *P. abies* and single *P. cembra* trees. There is no anthropogenic influence visible in the forest stand.

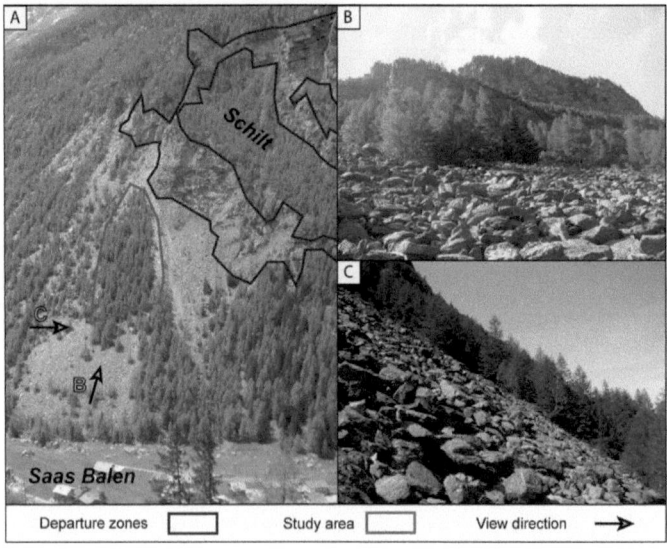

Fig. 2. (A). View of the study site and the departure zone. (B). View from to the cliffs generating rockfall. (C). Lateral view of the rockfall slope (mean slope angle 36°).

3 Materials and methods

3.1 Sampling strategy

Virtually all trees at the forest stand show clearly visible and severe growth disturbances (GD) caused by past rockfall events. On the study site, scars were the most prominent sign of previous rockfall activity. As the forest at Schilt has a protective function for the village of Saas Balen, only trees with a basal diameter <15 cm were felled. We investigated the entire upper forest stand and headed for an even distribution of trees throughout the study site. However, there is some concentration of sampled trees at the upper and lateral limits of the study area, as small trees predominantly grow in these sectors.

The positioning of the selected trees on the geomorphic map was based on aerial photographs and realized with a measuring tape, as the use of GPS devices was not possible in the dense forest stand within the deep valley.

Data recorded for each tree included microtopography or accumulation of rockfall deposits in the immediate vicinity of the tree. In addition, we noted tree-specific data including species, height, diameter at breast height, visible defects in its morphology such as e.g., scars (direction, surface and height of centre above ground level), broken crowns or branches, tilted stems or candelabra growth. We furthermore commented on neighboring trees if relevant.

Thereafter, photographs were taken of the entire tree and specific pictures for each wound so as to facilitate assigning GD to their triggering injury in the laboratory. Exclusively well-defined and clearly visible injuries were sampled at their maximum extension. In order to obtain a maximum of information, the sampling was realized using a handsaw, so as to obtain complete cross-sections from each injury.

3.2 Tree-ring analysis and intra-annual dating of events

The cross-sections were analyzed using standard dendrochronological methods (Stokes and Smiley, 1968; Bräker, 2002). Samples were first polished and tree rings counted, before tree-ring series were analyzed visually to identify GD caused by past rockfall.

The dating of past rockfall activity was based on the position of callus tissue and tangential rows of traumatic resin ducts (TRD) in *L. decidua* and *P. abies*, as both features are usually formed after cambium damage (Schweingruber, 1996, 2001). While callus tissue only occurs in the wood segment bordering the injury (Stoffel et al., 2005a, b; Per-ret et al., 2006), TRD can be even observed at some distance from the wound as well (Bollschweiler et al., 2007a; Stoffel, 2008). Bollschweiler et al. (2008) indicate a mean lateral TRD extension of 19% of the stem's circumference.

TRD were taken into account if they were present (i) in an extremely compact arrangement and (ii) forming continuous rows (Stoffel et al., 2005a). In this study, we did not focus on the appearance of compression wood, as it appears delayed and therefore is not suitable to date events with intra-annual precision (Timell, 1986). Additionally, in case of several wounds within a single year, it would have been impossible to assign the reaction wood to specific injuries.

Following Stoffel et al. (2005b), the intra-annual position of injuries and the adjacent callus tissue and TRD was determined as illustrated in Fig. 3 and impacts dated to the dormant season (D), early earlywood (EE), middle early-wood (ME), late earlywood (LE), early latewood (EL) and late latewood (LL). Based on data from neighboring sites (Müller, 1980; Stoffel et al., 2005b), we know that the vegetation period of *L. decidua*, *P. abies* and *P. cembra* locally starts at the end of May with the formation of thin-walled earlywood tracheids. The transition from LE to EL occurs in mid July and the formation of thick-walled latewood tracheids ends in early October. The period between October and May is called the "dormant season" (D) and there is no cytogenesis during this time of the year.

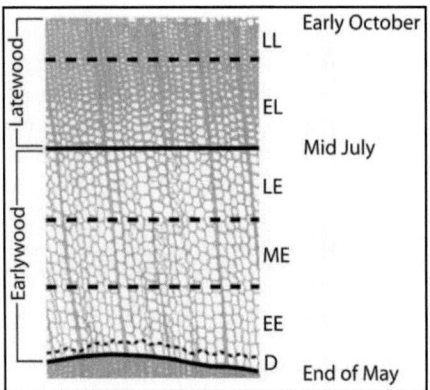

Fig. 3. *Subdivisions of a tree-ring: At the study site, earlywood formation lasts from end of May to mid July, latewood formation from mid July to early October. For the assessment of intra-annual rockfall activity, tree rings are further subdivided into early (EE), middle (ME) and late (LE) earlywood as well as early (EL) and late (LL) latewood (adapted from Stoffel et al., 2005b).*

While trees will react almost immediately to injuring events during the vegetation period (Stoffel, 2008), reactions to a rockfall impact caused during the dormant season will only be apparent at the very beginning of the "new" tree-ring.

3.3 Orientation of rockfall impacts

The position of the scar was assessed in the field with respect to the downslope direction. The idea behind this assessment was to approximate the source of rockfall material. Measurements were realized at the centre of each injury and the position of injuries noted in degrees (°). A frontal hit was attributed an impact angle of 0°, whereas 90° and 270° represent impacts located perpendicular to the general slope. Impact angles >90° and <270° stand for damage identified on the downslope part of the tree and would result from bouncing rocks and boulders leaving their marks over large parts of the tree's circumference. Results were grouped in classes of 30°.

4 Results

4.1 General aspects

Within this study, 32 trees were felled (23 *L. decidua*, 8 *P. abies*, 1 *P. cembra*) on the study site and 123 cross-sections prepared (Table 1), yielding a total of 154 injuries. Most frequently, one injury was identified per cross-sections, but multiple injuries were observed as well (26 cross-sections had 2 injuries, four cross-sections 3 injuries and one cross-section 4 injuries).

As can be seen from Table 2, age of the selected trees averaged 25.8 yrs (STDEV: 8.0 yrs), with the oldest one having 44 and the youngest 12 annual rings. Seven of the selected trees were present prior to 1975 whereas the other 25 individuals germinated between 1975 and 1995. The average diameter of all sampled trees is 9.1 cm (STDEV: 2.4 cm), representing a mean yearly diameter increase of 3.9 mm.

The nature of growth disturbances (GD) observed on the cross-sections following rockfall impacts is illustrated in Table 3. In total, 207 GD were identified on the cross-sections as a reaction to the 154 injuring impacts. TRD represented the most common reaction to impacts and were observed in 147 cases (95.5%). Interestingly, callus tissue was less frequently identified in the tree segments neighboring the injury and was only present on 60 cross-sections (39%, Table 3).

Trees sampled		32
	L. decidua	23
	P. abies	8
	P. cembra	1
Number of cross-sections		123
Number of injuries		154
	L. decidua	117
	P. abies	33
	P. cembra	4

Table 1. Number of trees, cross-sections and injuries analyzed.

	average	min.	max.	stdev
Tree age (in years)	25.8	12	44	7.5
Diameter (in cm)	9.1	5.4	15	2.4
Injuries per tree	4.8	1	16	3.2

Table 2. Tree age, tree diameter and number of injuries. Always indicating the average, the minimum value (min.), the maximum value (max.) and the standard deviation (stdev).

	Number	%
Total number of cross-sections	154	100
TRD	147	95.5
Callus growth	60	39

Table 3. Number of cross-sections and GD in absolute values and percentages.

4.2 Rockfall activity

A total number of 154 injuries were dated for the period 1975–2006, resulting in an overall mean activity of 4.8 hits yr^{-1}. Despite their predominantly young age and their rather small diameter, each tree shows a total number of 4.8 injuries in average.

The annual number of injuries as well as the sample depth (i.e. the number of trees present in a given year) is indicated in Fig. 4. It appears from the illustration that only five injuries occurred prior to 1985, when fewer than 75% of the sampled trees were present. This is why the seasonal timing or changes in rockfall activity will only be further analyzed for the period 1985–2006, when the number of trees is sufficient to obtain a record enough complete (≥ 25 trees). As a result, the mean rockfall activity augments to 6.8 hits yr^{-1}.

Figure 4 also indicates that there is a strong variation of rockfall activity between single years. Based on our data, rockfall activity was by far most important in 1995, when a total number of 66 wounds (43%) were recorded on the cross-sections, meaning that three out of four trees would have been injured. The second largest event was reconstructed for 2004, when a total of 23 injuries was observed on the cross-sections. In contrast, we also observe years with no injuries at all (1985, 1987 and 1989) and years with only one injury (1993 and 2000).

4.3 Seasonal timing of rockfall

The analysis of the intra-seasonal position of the 154 wounds and their neighboring callus tissue and TRD was used to determine the seasonal timing of rockfall activity at Schilt. Figure 5 clearly shows a concentration of

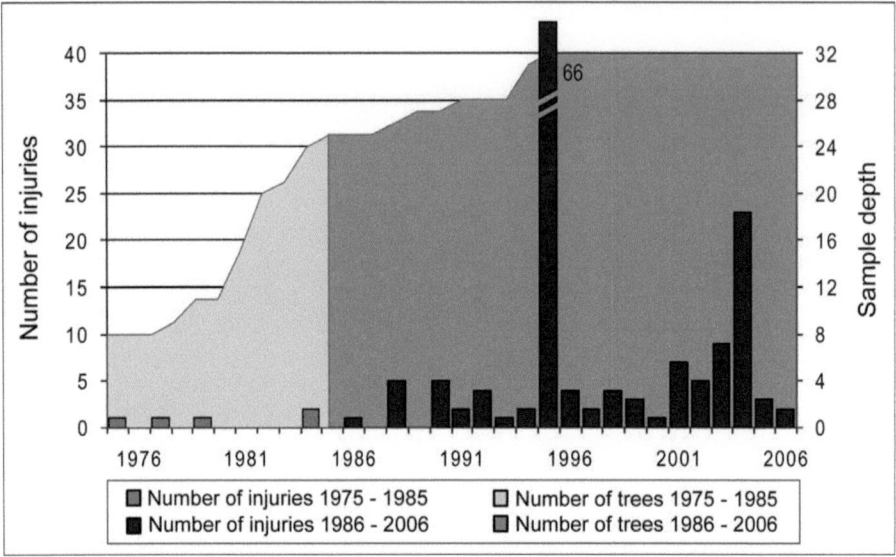

Fig. 4. Rockfall activity at Schilt as reconstructed with tree-ring analysis for the period 1975–2006. Due to the comparably young age of trees, analysis mainly focused on the period 1985–2006, when ≥75% of all trees were present (see sample depth).

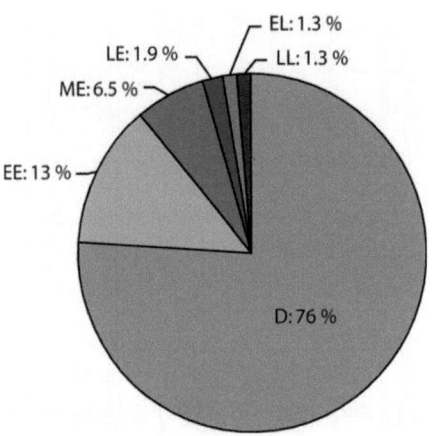

Fig. 5. Seasonal distribution of rockfall.

injuries that would have occurred during the dormant season of trees (76%) which locally lasts from early October to the end of May. Among the wounds located within the tree rings, it appears that rockfall is much more frequent during the period of earlywood formation (21.4%), i.e. from end of May to mid July – than during the period of latewood formation (only 2.6 %) i.e. between mid July and early October. Examples of injuries attributed to the dormant season as well as the periods of early- and latewood formation are provided in Fig. 6.

When going into further intra-seasonal detail for the early-wood events, it can be seen that the most significant activity (13%) occurred in EE, followed by 6.5% of the events dated to ME and only 1.9% to LE. The numbers further decrease when analyzing the latewood events, where both EL and LL register only 1.3% of all rockfall events.

In a subsequent step, the intra-annual behavior of rockfall was assessed for all years with ≥5 injuries. Table 4 provides an overview on the years that have been kept for further analysis. When analyzing the seaso-

vegetation period. Compared to the rockfall activity registered in the other years and when excluding the two major event years of 1995 and 2004, there is almost four times more activity observed in EE (23 vs. 6.2%) and two times more activity in ME (9.1 vs. 4.6%). Although rockfall was also important in 2004, the seasonal distribution of activity appears to be rather normal with just a slight concentration of events during D (87%) and only three events within the vegetation period (13%).

4.4 Orientation of rockfall impacts

The positioning of rockfall impacts with respect to the fall line (exact downslope direction) was noted in degrees (°) and results summarized in classes of 30° each. As can be seen from Fig. 7A, the average position of injuries with re-spect to the fall line was 13.3° and most wounds were at-tributed to the classes 16–45° (24%), 46–75° (14.9%) and 345–15° (8.2%). As expected, only eight injuries (5%) were observed on the downslope part of the stem (90–270°), probably resulting from deviated rocks and boulders. As it can be seen in illustration 7B, directional results are consistent with the geomorphic situation in the field, where the rockfall generating cliffs are located in the direction the average position of the injuries is pointing at.

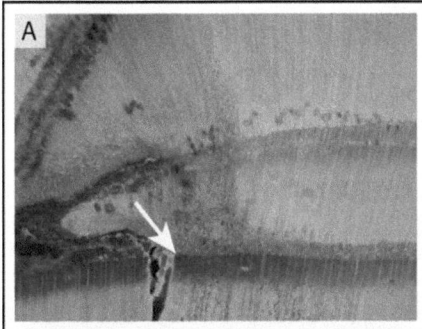
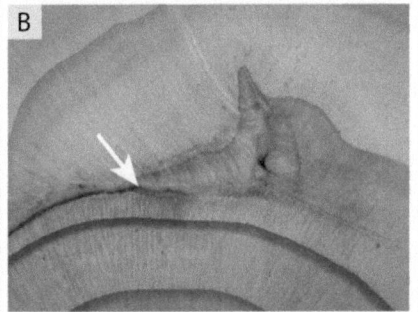
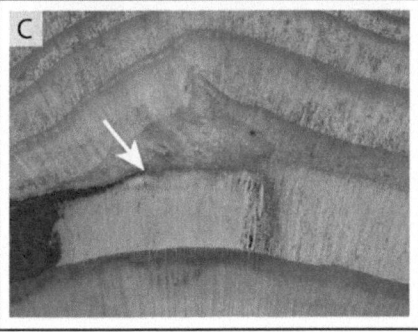

Fig. 6. Examples of rockfall injuries located at different positions within the tree ring:. Injury attributed to (A) the dormant season, (B) mid earlywood and (C) early latewood.

nal behavior of rockfall in 1995, it appears that activity was not only exceptionally high, but also different with respect to the seasonal timing of injuring events. As can be seen from Table 4, a comparably high number of impacts apparently occurred outside the dormant season in the early stages of the new

5 Discussion

In the study we report here, 123 cross-sections from 32 trees (*Larix decidua* Mill., *Picea abies* (L.) Karst., *Pinus cembra* L.) were used to reconstruct yearly rockfall activity, the seasonal behavior of rockfall as well as to determine the main direction and the source areas of rockfall for the last 32

	D	%	EE	%	ME	%	LE	%	EL	%	LL	%
∑	*117*	*76*	*20*	*13*	*10*	*6.5*	*3*	*1.9*	*2*	*1.3*	*2*	*1.3*
∑ without 1995/2004	52	80	4	6.2	3	4.6	3	4.6	2	3.1	1	1.5
2004	20	87	1	4.3	1	4.3	0	0	0	0	1	4.3
2003	6	86	1	14	0	0	0	0	0	0	0	0
2002	5	100	0	0	0	0	0	0	0	0	0	0
2001	5	83	0	0	0	0	0	0	0	0	1	17
1995	45	68*	15	23*	6	9.1*	0	0	0	0	0	0
1990	4	80	0	0	0	0	0	0	1	20	0	0

Table 4. Intra-seasonal distribution of rockfall activity at Schilt, Saas Balen. The overall distribution of rockfall (6) is given in italics and rockfall activity illustrated for those years with >5 injuries. The abnormal seasonal distribution of rockfall activity in 1995 is highlighted with *.

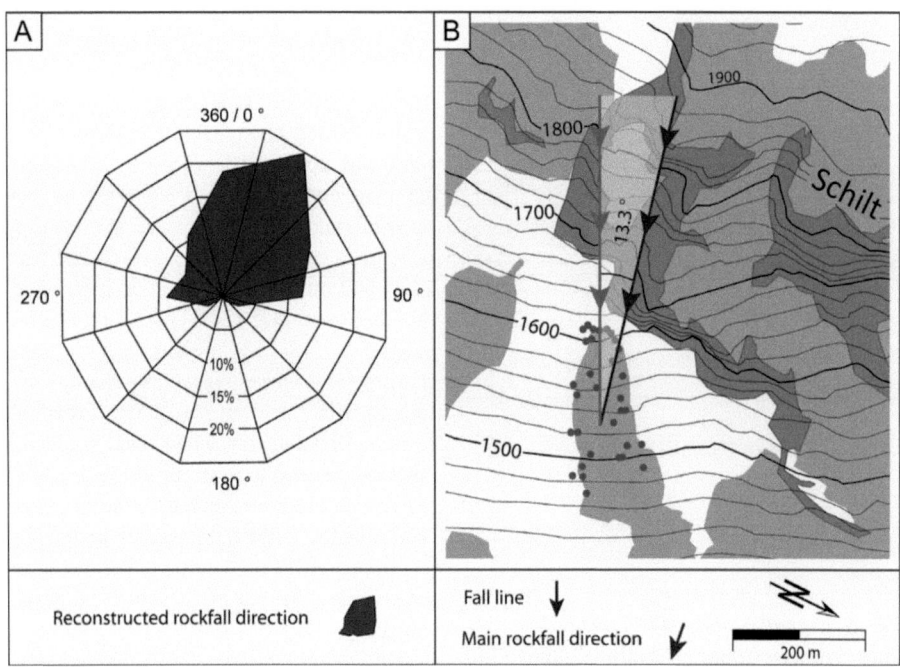

Fig. 7. (A). Radial distribution of rockfall impacts as observed on the selected trees. (B). The average rockfall direction (red) diverges from the fall line and points toward the main source area.

years. In total, 154 injuries were identified that have induced a total of 207 growth disturbances.

Within this study, we exclusively worked with cross-sections from trees with a DBH <15 cm. In contrast to the analysis of increment cores where TRD can occur in later segments of the tree ring with increasing axial or tangential distance from the injury (see Bollschweiler et al., 2008), the analysis of rockfall scars on cross-sections and the identification of reactions in the tissues bordering the wound allowed an intra-annual dating with very high accuracy. The felling of comparably small trees resulted in a rather young age of samples (25.8 yrs).

Our results suggest that the formation of TRD is the most common and widespread reaction of *L. decidua* and *P. abies* to rockfall and was present on 95.5% of the samples. TRD can therefore be used as an excellent marker of past rockfall events, even more as no trees were showing TRD without the presence of a nearby injury. Callus tissue could, in contrast, only be assessed on 39% of the samples and was generally much less widespread within the tree ring than TRD. As our data are exclusively based on young trees and on a comparably small number of cross-sections, further research is needed to study the formation of these features following rockfall impacts.

As for the rockfall activity in individual years, the reconstruction revealed strong inter-annual differences, with major activity observed in 1995 and 2004 and other years with no events at all observed in the tree-ring records. These results correlate with findings of Stoffel et al. (2005b) or Perret et al. (2006), who both describe distinct inter-annual fluctuations of rockfall activity as well on their sites located in the Valais Alps and Bernese Oberland, respectively.

The clustering of missing injuries for several years in the early 1980s can be explained by the small number of sampled trees alive for that time, the smaller diameter and the younger age of the individuals (more flexible and therefore probably less vulnerable). It is therefore worthwhile to note that the reconstructed number of events does not represent a complete frequency reconstruction, as relevant factors (e.g. regular arrangement of sampled trees or varying tree diameters) have not been taken into consideration. The probability of an impact is therefore not evenly distributed, neither in space, nor in time. In a similar way, no clear trend can be deduced from the rockfall series we have obtained.

The intra-annual dating of injuries indicates that rockfall at Schilt predominantly occurred during the dormant season which locally lasts from early October to the end of May (76%). Our results are in very close agreement with the findings of Perret et al. (2006) in the Diemtigtal (Swiss Prealps), reporting 74% of rockfall activity during the dormant season. In contrast, Stoffel at al. (2005b) dated a higher percentage of injuries to the dormant season (88%) in the Matter Valley (Swiss Alps). It is most probable that the difference in the intra-annual activity of rockfall is due to the fact that there is contemporary permafrost present in the departure zone at the site studied by Stoffel et al. (2005b), but not in the cliffs of our site or in that investigated by Perret et al. (2006). Despite the concentration of rockfall activity to the dormant season, the comparably low values identified for the period of earlywood formation should not be underestimated, as more than 20% of the annual rockfall activity apparently takes places in less than two months (i.e. between the end of May and mid July). It is feasible that this concentration of rockfall activity would be the result of the annual thaw of winter ice and the circulation of melt water in preexisting fissures (Matsuoka, 2006; Hall, 2007).

Another similarity between the reconstructed data from our study site and those presented by Stoffel et al. (2005b) on rockfall activity on a slope in the neighboring Matter Valley (see Fig. 1A) is that rockfall activity was exceptionally high at both sites in 1995.

The massive concentration of rockfall activity in 1995 allows a more detailed seasonal analysis of rockfall activity within the year, indicating that there was three times more activity occurring during EE and ME as compared to the mean distribution. Apart from occasional freeze-thaw cycles or the melting of ice, it is possible that heavy precipitation events could have led to this concentrated and massive re lease of rocks and boulders. Meteorological data from a station in Zermatt (located 20 km to the south-west of the study site; SMI 2007) indicate indeed exceptionally heavy precipitation events for the dormant season of 1994–1995, with the heaviest daily precipitation sum for at least the last 25 years recorded on 5 November 1994 (Fig. 8). As the local meteorological station of Saas Balen is only operational since December 1994, we do not have any indications on the intensity of this precipitation event at the study site. Similarly and based on the air temperatures measured at Zermatt, it is not quite clear whether the precipitation at our study site was in the form of rain or snow. In contrast, the local meteorological station recorded the second most important precipitation event between 1994 and today on 21 April 1995. As the two rockfall sites in the two neighboring valleys (Stoffel et al., 2005b, our study site) presumably reacted to the same triggering events, we suppose that the concentration of several precipitation events affecting large parts of the Valais Alps would be at the origin of the much higher frequency of rockfall in the first half of 1995.

We also have to admit that the response of the cliffs in generating rockfall following the precipitation may appear to be delayed.

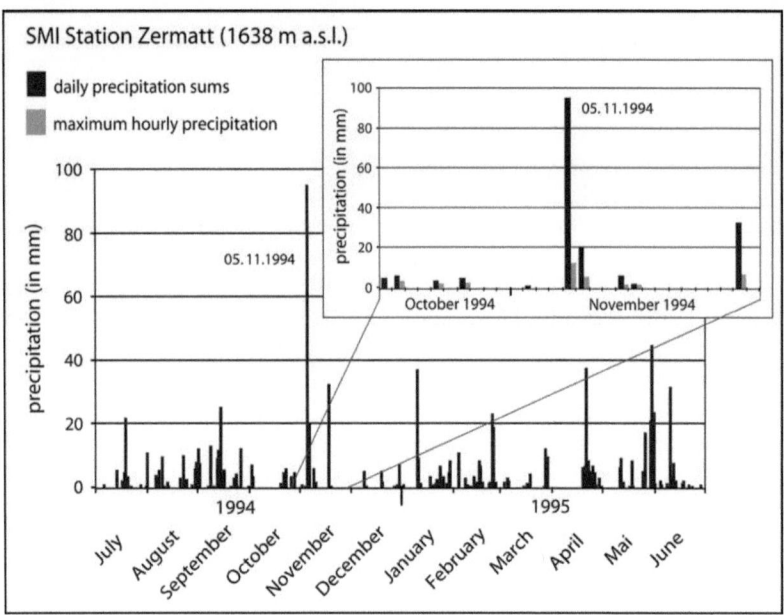

Fig. 8. Meteorological data from the SMI station Zermatt confirm exceptionally high daily precipitation (in blue) recorded on 5th November 1994 with 94.4 mm/24h. This represents the most intense rainfall for at least 25 years. The hourly maximum attained during this event was 11.3 mm (in red), indicating a persistent intense rainfall rather than an extreme and short thunderstorm.

There are two possible explanations for this delay: Air temperatures may have dropped suddenly after the heavy rainfalls in November 1994 and/or April 1995, leading to abnormally large quantities of water frozen in the cracks. On the other hand, it is feasible that snowfall occurred during both events, leading to large quantities of water flowing through the cracks during snow melt.

Data obtained on the position of scars on the stem surface and the resulting back tracing of "rockfall trajectories" are in very good agreement with the geomorphic situation in the field, as the main direction of rockfall pointed towards the main rockfall source area. We are well aware that falling rocks and boulders may be deviated during ground contact and by impacting other trees further up on the slope, nevertheless results indicate that some directional information was preserved.

6 Conclusions

Tree-ring analysis allowed inter-annual dating of 154 past rockfall injuries between 1975 and 2006 as well as the assessment of intra-seasonal behavior of rockfall at the study site. We conclude that the tree-ring analysis delivers exceptionally good data on the inter- as well as intra-annual behavior of rockfall and that it furnishes valuable information on the activity of rockfall source areas. At the same time and as rockfall activity peaks during the dormant season of trees, it is not possible to date these events with higher temporal resolution.

Acknowledgements

The authors acknowledge the valuable comments from the reviewers D. R. Butler and J. Moya. They are also grateful to M. Bollschweiler for her assistance in the field and lab and for the many insightful comments on earlier versions of this paper. Furthermore we would like to thank L. Correa for his assistance in the field and O. Hitz for his valuable comments on wood anatomy. Finally, we kindly thank the local administration and the forester for allowing us to work in their forest.

References

Azzoni, A., Barbera, G. L., and Zaninetti, A.: Analysis and prediction of rockfalls using a mathematical model, Int. J. Rock. Mech. Min., 32, 709–24, 1995.

Bearth, P.: Geologischer Atlas der Schweiz 1:25000, Niklaus, (Atlasblatt 71), Schweizerische Geologische Kommission, 1980.

Bearth, P.: Geologischer Atlas der Schweiz 1:25000, Simplon, (Atlasblatt 61), Schweizerische Geologische Kommission, 1973.

Berger, F., Quetel, C., and Dorren, L. K. A.: Forest: A natural protection mean against rockfall, but with which efficiency? The objectives and methodology of the ROCKFOR project, Interpavement 2002, 2, 815–826, 2002.

Bollschweiler, M., Stoffel, M., Schneuwly, D. M., and Bourqui, K.: Traumatic resin ducts in *Larix decidua* stems impacted by debris flows, Tree Physiol., 28, 255–263, 2008.

Bozzolo, D., Pamini, R., and Hutter, K.: Rockfall analysis – a math-ematical model and its test with field data, Proceedings of the 5th International Symposium on Landslides in Lausanne, Rotterdam, Balkema, 555–60, 1986.

Braam, R. R., Weiss, E. E. J., and Burrough, P. A.: Spatial and temporal analysis of mass movement using dendrochronology, Catena, 14, 573–584, 1987.

Bräker, O. U.: Measuring and data processing in tree-ring research – a methodological introduction, Dendrochronologia, 20, 203– 216, 2002.

Butler, D. R.: Rockfall hazard inventory, Ram River, Mackenzie Mountains, The Canadian Geographer, 27, 2, 175–178, 1983.

Coe, J. A. and Harp, E. L.: Influence of tectonic folding on rockfall susceptibility, American Fork Canyon, Utah, USA, Nat. Hazards Earth Syst. Sci., 7, 1–14, 2007, http://www.nat-hazards-earth-syst-sci.net/7/1/2007/.

Davies, M. C. R., Hamza, O., and Harris, C.: The effect of rise in mean annual temperature on the stability of rock slopes containing ice-filled discontinuities, Permafrost Periglac., 12, 1, 137– 144, 2001.

Dorren, L. K. A., Berger, F., and Putters, U. S.: Real size experiments and 3D simulation of rockfall on forested and non-forested slopes, Nat. Hazards Earth Syst. Sci., 6, 145–153, 2006, http://www.nat-hazards-earth-syst-sci.net/6/145/2006/.

Douglas, G. R.: Magnitude frequency study of rockfall in Co. Antrim, N. Ireland, Earth Surf. Proc. Land., 5, 2, 123–129, 1980.

Erismann, T. H.: Flowing, rolling, bouncing, sliding, synopsis of basic mechanisms, Acta Mech., 64, 101–110, 1986.

Evans, S. G. and Hungr, O.: The assessment of rockfall hazard at the base of talus slopes, Can. Geotech. J., 30, 620–36, 1993.

Gardner, J.: Rockfall frequency and distribution in the Highwood Pass area, Canadian Rocky Mountains, Z. Geomorphol., 27, 311–24, 1983.

Gardner, J. S.: Frequency, magnitude, and spatial distribution of mountain rockfalls and rockslides in the Highwood Pass Area, Alberta, Canada, in: Thresholds in Geomorphology, edited by: Coates, R., Vitek, J. D., Allen and Unwin, New York, 267– 295, 1980.

Gruber, S., Hoelzle, M., and Haeberli, W.: Permafrost thaw and destabilization of Alpine rock walls in the hot summer of 2003, Geophys. Res. Lett., 31, L13504, 2004.

Guzzetti, F., Crosta, G., Detti, R., and Agliardi, F.: STONE: a com-puter program for the three-dimensional simulation of rock-falls, Computers and Geosciences, 28, 9, 1079–1093, 2002.

Hall, K.: Evidence for freeze-thaw events and their implications for rock weathering in northern Canada. II. The temperature at which water freezes in rock, Earth Surf. Proc. Land., 32, 2, 249– 259, 2007.

Harp, E. L. and Wilson, R. C.: Shaking intensity thresholds for rock falls and slides: Evidence from 1987 Whittier Narrows and superstition hills earthquake strong-motion records, B. Seismol. Soc. Am., 85, 6, 1739–1757, 1995.

Hungr, O. and Evans, S. G.: Engineering evaluation of fragmental rockfall hazards, Proceedings of the 5th International Sympo-sium on Landslides in Lausanne, Rotterdam, Balkema, 685–90, 1988.

Kirkby, M. J. and Statham, I.: Surface stone movement and scree formation, J. Geol., 83, 349–62, 1975.

Lafortune, M., Filion, L., and Hétu, B.: Dynamique d'un front forestier sur un talus d'éboulis actif en climat tempéré froid (Gaspésie, Québec), Géogr. Phys. Quat, 51, 1, 1–15, 1997.

Luckman, B. H.: Rockfalls and rockfall inventory data; some observations from the Surprise Valley, Jasper National Park, Canada, Earth Surf. Proc. Land., 1, 287–98, 1976.

Luckman, B. H. and Fiske, C. J.: Estimating long-term rockfall accretion rates by lichenometry, Slaymaker, O. (Ed.), Steepland Geomorphology, Wiley, Chichester, 233– 255, 1995.

Marzorati, S., Luzi, L., and De Amicis, M.: Rock falls induced by earthquakes: a statistical approach,

Soil Dyn. Earthq. Eng., 22, 7, 565–577, 13, 2002.

Matsuoka, N.: Frost wedging and rockfalls on high mountain rock slopes: 11 years of observations in the Swiss Alps, Geophys. Res. Abstr., 8, 05344, European Geosciences Union, 2006.

Matsuoka, N. and Sakai, H.: Rockfall activity from an alpine cliff during thawing periods, Geomorphology, 28, 3, 309–328, 1999.

McCarroll, D., Shakesby, R. A., and Matthews, J. S.: Spatial and temporal patterns of Late Holocene rockfall activity on a Norwegian talus slope: lichenometry and simulation-modelling approach, Arct. Alp. Res., 30, 51–60, 1998.

Müller, H.-N.: Jahrringwachstum und Klimafaktoren: Beziehungen zwischen Jahrringwachstum von Nadelbaumarten und Klimafaktoren an verschiedenen Standorten im Gebiet des Simplonpasses (Wallis, Schweiz), Forstl. Bundes-Versuchsanst., Wien 25, Agrarverlag, Wien, p. 81, 1980.

Okura, Y., Kitahara, H., Sammori, T., and Kawanami, A.: The effects of rockfall volume on runout distance, Eng. Geol., 58, 2, 109–124(16), 2000.

Perret, S., Stoffel, M., and Kienholz, H.: Spatial and temporal rockfall activity in a forest stand in the Swiss Prealps – a dendrogeomorphological case study, Geomorphology, 74, 219–231, 2006.

Ritchie, A. M.: Evaluation of rockfall and its control, Washington, DC: Highway Research Board, National Research Council, Highway Research Record, 17, 13–28, 1963.

Ruppen, P. J., Imseng, G., and Imseng, W.: Saaser Chronik 1200–1979, Rotten-Verlag, Brig, p. 54, 1979.

Sass, O.: Temporal Variability of Rockfall in the Bavarian Alps, Germany, Arct. Antarct. Alp. Res., 37, 4, 564–573, 2005.

Schweingruber, F. H.: Tree Rings and Environment, Dendroecology, Paul Haupt, Bern, p. 82, 1996.

Schweingruber, F. H.: Holzanatomie, Paul Haupt, Bern, 334–345, 2001.

Shroder, J. F.: Dendrogeomorphology: Review and new techniques of tree-ring dating, Prog. Phys. Geog., 4, 161–188, 1980.

SMI (Swiss Meteorological Institute): Annals of the Swiss Meteorological Institute, daily precipitation sums 1982–2007, Zurich, http://www.sma.ch, 2007.

Solomina, O. N.: Dendrogeomorphology: Research Requirements, Dendrochronologia, 20, 1, 231–243, 2002.

Statham, I. and Francis, S. C.: Influence of scree accumulation and weathering on the development of steep mountain slopes, Abrahams, A. D. (Ed.), Hillslope processes, Winchester, Allen and Unwin Inc., 245–67, 1986.

Stoffel, M., Schneuwly, D., Bollschweiler, M., Lièvre, I., Delaloye, R., Myint, M., and Monbaron, M.: Analyzing rockfall activity (1600–2002) in a protection forest – a case study using dendrogeomorphology, Geomorphology, 68, (3–4), 224–241, 2005a.

Stoffel, M., Lièvre, I., Monbaron, M., and Perret, S.: Seasonal timing of rockfall activity on a forested slope at Täschgufer (Valais, Swiss Alps) – a dendrochronological approach, Z. Geomorphologie, 49, 1, 89–106, 2005b.

Stoffel, M., Bollschweiler, M., and Hassler, G.-R.: Differentiating past events on a cone influenced by debris-flow and snow avalanche activity – a dendrogeomorphological approach, Earth Surf. Proc. Land., 31, 1424–1437, 2006a.

Stoffel, M., Wehrli, A., Kühne, R., Dorren, L. K. A., Perret, S., and Kienholz, H.: Assessing the protective effect of mountain forests against rockfall using a 3D simulation model, Forest. Ecol. Manag., 225, 113–122, 2006b.

Stoffel, M., Conus, D., Grichting, M. A., Lièvre, I., and Maître, G.: Unraveling the patterns of late Holocene debris-flow activity on a cone in the Swiss Alps: Chronology, environment and implications for the future, Global Planetary Change, 60, 222–234, 2008.

Stoffel, M.: Dating past geomorphic processes with tangential rows of traumatic resin ducts, Dendrochronologia, 26, 53–60, 2008.

Stokes, M. A. and Smiley, T. L.: An Introduction to Tree-Ring Dat-ing, Chicago, University of Chicago Press, 73 pp., 1968.

Timell, T. E.: Compression wood in gymnosperms, Springer-Verlag, Berlin, 2150 pp., 1986.

Wiles, G. C., Calkin, P. E., and Jacoby, G. C.: Tree-ring analysis and Quaternary geology: Principles and recent applications, Ge-omorphology, 16, 259–272, 1996.

♦♦♦♦♦

published in Geomorphology 2008, Vol. 102, 522–531

Spatial analysis of rockfall activity, bounce heights and geomorphic changes over the last 50 years – A case study using dendrogeomorphology

Dominique M. Schneuwly and Markus Stoffel

Laboratory of Dendrogeomorphology, Department of Geosciences, University of Fribourg, Chemin du Musée 4, 1700 Fribourg, Switzerland

Abstract

Tree-ring series have been used to reconstruct 50 years of rockfall behavior on a slope near Saas Balen (Swiss Alps). A total of 796 cores and 141 cross sections from 191 severely injured conifer trees (*Larix decidua* Mill., *Picea abies* (L.) Karst. and *Pinus cembra* L.), combined with a series of aerial photographs, were used to investigate the evolution of the forest stand so as to (i) to reconstruct past rockfall rates; (ii) to analyze the spatial behavior of maximum bounce heights; and (iii) to analyze the spatial comportment of rockfall activity over the last five decades.

Tree-ring analysis permitted the reconstruction of the age distribution at the study site; results were in perfect agreement with the afforestation process shown in the aerial photographs. The oldest are located in the lower, central part of the study site; the youngest individuals at its uppermost lateral boundaries. Reconstructed rockfall rates reveal strong interannual variations and single event years with increased activity, namely in 1960/1961 and 1995. Spatial analysis of the maximum bounce heights indicate highest values at the lateral boundaries and lowest heights in the lower central part of the forest stand, where a big boulder seems to shield trees growing below it. The spatial analysis of past rockfall activity shows high active zones at the uppermost north-facing boundaries of the forest and least active zones in the lowermost central part of the studied stand. The high rockfall activity at the slope is expressed by a mean rockfall rate of > event $m^{-1} y^{-1}$.

1. Introduction

Rockfall is one of the most common mass movement processes in mountain regions and is defined as the free falling, bouncing or rolling of individual or a few rocks and boulders, with volumes involved generally being b5 m3 (Berger et al., 2002). As rockfall potentially endangers humans and infrastructure, it has become one of the most intensely studied geomorphic processes of the cliff zone in mountain areas.

Therefore, numerous studies exist concerning various aspects of rockfall, such as the dynamic behavior (Ritchie, 1963; Erismann, 1986; Azzoni et al., 1995), boulder reaction during ground contact (Bozzolo and Pamini, 1986; Hungr and Evans, 1988; Evans and Hungr, 1993), or runout distances of falling rocks (Kirkby and Statham, 1975; Statham and Francis, 1986; Okura et al., 2000). Much research was also done on the possible triggers of rockfall, such as freeze-thaw cycles (Gardner, 1983; Matsuoka and Sakai, 1999; Matsuoka, 2006), changes in the rock-moisture level (Sass, 2005), the thawing of permafrost (Gruber et al., 2004), the increase of mean annual temperatures (Davies et al., 2001), tectonic folding (Coe and Harp, 2007) or the occurrence of earthquakes (Harp and Wilson, 1995; Marzorati et al., 2002). In addition, several studies exist on the long-term accretion rates of rockfall (Luckman and Fiske, 1995; McCarroll et al., 1998). Furthermore, since the late 1980s, the field of numeric modelling has become a major topic in the field of rockfall research (Zinggeler, 1989; Guzzetti et al., 2002; Dorren et al., 2006; Stoffel et al., 2006).

However, there are only a few studies that analyze the temporal evolution of rockfall. Some observation-based studies were conducted over one or several summers (Luckman, 1976; Douglas, 1980; Gardner, 1980), yielding valuable data on the contemporary activity, but major difficulties were encountered in estimating long-term changes in rockfall accretion rates. Douglas (1980), for example, performed detailed observations of rockfall in Co. Antrim, Ireland, on a weekly basis and over a period of two years. Results indicate seasonal peaks of rockfall activity in February and March, as well as in November and December, and a constant but small background activity during the rest of the year. More recently, Sass (2005) installed more than 60 rockfall barriers in the Bavarian Alps so as to quantify rockfall activity over a period of four years. He concluded that the distribution of rockfall is highly variable in time and that it is triggered by a combination of a multitude of parameters.

While these *in situ* observations provide detailed data sets on short periods of time, they are, however, too limited for revealing trends or years with above-average activity. Luckman and Fiske (1995) used lichens to estimate rockfall rates for the last 300 years, but could only deliver a spatial resolution of 50 years; while McCarroll et al. (1998) estimated rockfall rates for the late Holocene with a temporal resolution of 25 years using lichenometry.

In a similar way, only a few studies exist that reconstruct the spatial behavior of rockfall on a slope. Gardner (1980) used a spatial grid of 500 m during his fieldwork, rendering detailed spatial analysis impossible. Luckman and Fiske (1995) worked on two different sectors of one large talus slope, but did not look at spatial variations between or within the areas.

In contrast to other mass movement processes (Shroder, 1978, 1980; Wiles et al., 1996; Solomina, 2002), tree-ring analyses have only rarely been used in rockfall research. Lafortune et al. (1997) reconstructed sedimentation rates on a scree slope by analy-

zing tree-rings. Several recent studies have been conducted using tree-ring analyses so as to investigate different rockfall aspects, namely in assessing seasonal timing (Stoffel et al., 2005a; Schneuwly and Stoffel, 2008) or to investigate the evolution of rockfall over time (Stoffel et al., 2005b). Stoffel et al. (2005b) used a new approach to investigate combined spatial and temporal variations of rockfall. They analyzed 135 trees severely injured by rockfall to reconstruct historic rockfall rate with its spatial behavior. As the ~800 growth anomalies covered four entire centuries (1600–2002), the rockfall rate had to be reconstructed on a decadal basis to provide a reliable data set. In addition, the spatial distribution is not necessarily representative of the present-day behavior of rockfall activity at the study site.

It is therefore the aim of this study to use dendrogeomorphological methods to (i) investigate the evolution of the Schilt forest in the Valais Alps (Switzerland); (ii) determine yearly rockfall rates; and (iii) analyze the spatial behavior of rockfall aspects. Results were obtained from 937 samples of 191 European larch (*Larix decidua* Mill.), Norway spruce (*Picea abies* (L.) Karst.), and Swiss stone pine (*Pinus cembra* L.) trees on an homogenous rockfall slope, thus providing a very dense data set of more than 2050 growth disturbances resulting from rockfall activity over the last five decades.

2. Study area

The study site is situated in the "Schilt" forest near Saas Balen (46°09'06" N., 7°55'27" E.) in the Valais Alps, Switzerland (Fig. 1). The forest stand investigated is located between 1390 and 1610 m asl on the ENE-facing slope below the Lammenhorn (3189 m asl). The bedrock in the zone of origin consists

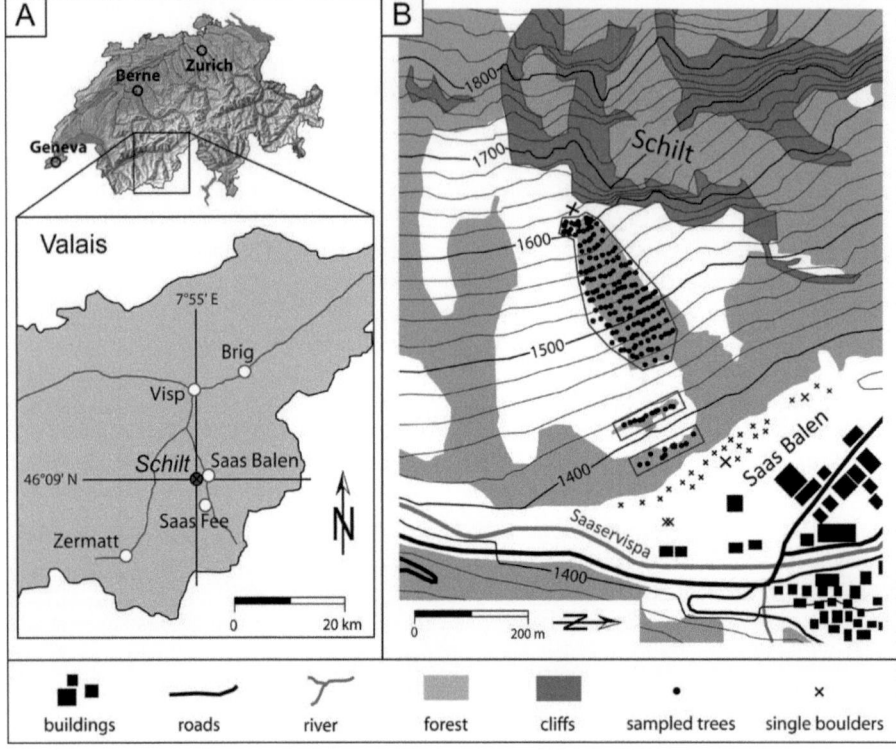

Fig. 1. (A) The study site is located in the Saas Valley southwest of Saas Balen. (B) Sketch map of the study site with the position of the trees sampled, the boulders are in the valley bottom and the large block (2000 m³) is at the top of the study site.

of micaceous schists belonging to Penninic crystalline layers, dipping SSE with an angle of ca. 20° (Bearth, 1973, 1980). Rockfall frequently occurs on the slope, originating from the disintegrated and glacially oversteepened cliffs at 1750–1900 m asl (Fig. 2). Quaternary talus and morainic deposits cover the transition and deposition zones.

Archival data and local toponomy indicate the occurrence of rockfall in the region since at least the early eighteenth century when rockfalls descended from the neighboring "Steinschlagwald" (=rock-fall forest) and destroyed the old parish church (Ruppen et al.,1979). In contrast, other mass movement processes, such as debris flows or snow avalanches, have never been witnessed on the slope.

The volume of falling rocks does not normally exceed 1 m³ on the study site. However, few blocks are deposited in the valley bottom with volumes of up to 50 m³, bearing witness to major events in the past. In addition, the largest block on the entire slope, unique in its kind, is located at the uppermost end of the investigated area and has an estimated volume of 2000 m³.

The study site is 300 m long and has an average width of 75 m, covering a surface of ~20,000 m². The mean slope angle is 36° with only a small variation between the top

Fig. 2. (A) View of the study area and the origin zones. (B) View towards the cliffs generating the rockfall. (C) Lateral view of the rockfall slope (mean slope angle 36°). (D) Big boulders in the valley bottom bearing witness to major Holocene rockfall events.

and the bottom of the study site. A centimetric soil layer covers rockfall deposits within the study site, whereas the surfaces outside the study area remain almost free of vegetation and are covered with bare rocks and boulders. The stand at Schilt mainly consists of *Larix decidua* Mill. trees (90%), of some *Picea abies* (L.) Karst. (10%), and single *Pinus cembra* L. (<1%). Three separate forest units exist; the largest located in the upper half of the site is covered with a stand that becomes gradually denser towards its upper limits. The two other units each consist of small forested bands oriented perpendicular to the fall line and are located at the bottom of the investigated area. Between the largest unit and the first band there is a distance of 60 m. The gap between the first (1390 m asl) and the second forest band (1410 m asl) is 40 m. No anthropogenic influence is visible in the forest stand, either at the study site itself or in neighboring forested sectors.

3. Material and methods

3.1. Sampling strategy

At the study site, virtually all of the trees show clearly visible and severe growth disturbances (GD) caused by past rockfall events in the form of scars, broken crowns or branches, and tilted stems, with scars being the most prominent sign of previous rockfall activity in the predominantly young trees. As scars represent the most accurate and reliable GD to date past rockfall events (Schneuwly and Stoffel, 2008), we focused sampling exclusively on the analysis of wounds. Sampling was performed along horizontal transects in order to investigate the entire stand and to assure an even distribution of trees throughout the study site. Transects were chosen every 15 m, and one tree was selected every 10 m along each transect. As the use of GPS devices was not possible with precision in the steep valley and within the forest at "Schilt", the positioning of the selected trees on the map was based on aerial photographs and using a measuring tape.

Sampling of trees was undertaken using two different methods: (i) on one hand, we used a small handsaw to obtain entire cross sections from each injury of trees with a basal diameter <15 cm; (ii) on the other hand, we extracted cores (max. 40 cm×5 mm) with increment borers. Injuries in bigger trees (>15 cm) were sampled with at least one core per scar, with the core extracted from the overgrowing tissue at a height showing the maximum wound extension. As the extraction of cores from the overgrowing callus needs to be effected at the contact of the injured with the non injured tissue, normally more than one sample had to be taken to obtain adequate core samples.

In addition, data recorded for each tree included micro topography or the accumulation of rockfall deposits in the immediate vicinity of the tree. We also noted tree-specific data including specie, height, diameter at breast height (DBH), and comments on neighboring trees. All visible defects in its morphology (such as scars (direction, surface and height of centre above ground level), broken crowns or branches, and tilted stems or candelabra growth) were recorded and general pictures of the tree and for each wound were taken to facilitate rockfall reconstruction in the laboratory. In a final step and in case any increment cores were taken, we noted the height and direction of each core sample.

3.2. Tree-ring analysis

The disturbed tree samples were analyzed using standard dendrochronological methods (Stokes and Smiley, 1968; Bräker, 2002). The samples were first polished and the tree rings counted, before tree-ring series were analyzed visually to identify GD caused by past rockfall. Among the different reactions, we focused on the occurrence of tangential rows of traumatic resin ducts (TRD; Bollschweiler et al., 2008; Schneuwly and Stoffel, 2008; Stoffel, 2008), the presence of callus tissue (Schweingruber, 2001), eccentric growth and the formation of compression wood (Braam et al., 1987a; Fantucci and Sorriso-Valvo, 1999), as well as abrupt changes in growth (Strunk, 1997; Friedman et al., 2005).

Within this study, we primarily focused on the formation of resin ducts, although they may have causes other than rockfall, e.g., climatic stress, high winds, insect attack, or fraying and cropping by ungulates. However, as sampling was performed exclusively next to visible wounds, we can be certain that all TRD present on the cores were the result of disturbances caused by falling rocks. Nevertheless, to assure a well-defined classification, we applied the limitations defined by Stoffel et al. (2005b), who only considered resin ducts to be the result of rockfall activity if they were (i) traumatic, (ii) extremely compact, and (iii) forming continuous rows.

Callus tissue is formed after cambium damage (Schweingruber, 1996, 2001). It occurs primarily at the borders of rockfall injuries (Stoffel et al., 2005a,b; Perret et al., 2006) and had to be clearly present to be taken into account. In case several cores of a single tree showed TRD in the same year, they were considered as one single event, as the injuries could have been caused by the same event or even the same rock. In case several TRD were present on a single core in different years, the latter were only considered if there were thought to be at least five undisturbed years between the two occurrences. Consequently, results presented in this study have to be considered as minimum frequencies.

Following the ideas of Braam et al. (1987b), reaction wood was only considered if the growth change was sudden and over a period of five or more years. Finally, ring widths were analyzed visually so as to identify abrupt changes in growth following impacts (decrease or increase; Schweingruber, 1996, 2001).

In addition, the chance of identifying rockfall scars in trees increases with a rising number of trees on one hand and with the continuous diameter growth of trees on the other hand. In this sense, a thick stem exposes a larger target to falling rocks and is more easily struck than a smaller one. Per Stoffel et al. (2005b), we therefore used a rockfall "rate" expressed as the number of rockfall events per meter width of all tree surfaces (DBH sum) present per year so as to be able to compare present with past rockfall activity. To assess the annual rockfall rate, the DBH increment of every individual tree per year was determined by dividing its DBH by the number of rings between pith and sample year (2006) at breast height. The DBH values of all trees existing at the beginning of a particular year were then summarized to include what we refer to as exposed diameter (ED; in m). To obtain the rockfall "rate", the yearly sum of reconstructed events was finally divided by the ED, indicating the number of events recorded per meter ED and per year.

In order to investigate the spatial behavior of rockfall activity, the rockfall rate at the position of each tree had to be calculated. Therefore, the average DBH of each individual was assessed as a first step by dividing its DBH by two, resulting in a simplified mean ED per tree. This mean ED was then multiplied by the age of the tree, producing the cumulative ED of each tree. Finally, the number of events per tree was divided by its cumulative ED so as to calculate the number of events $m^{-1} y^{-1}$.

3.3. Spatial visualization of tree-ring data

The general evolution of the forest stand at the entire study site was first investigated with six aerial photographs taken in different years (1941, 1958, 1968, 1981, 1993, and 2004). The pictures were initially georeferenced and then visually analyzed.

However, only the large stand in the upper half of the study site was investigated in detail with geographic information systems GIS as the spatial relationship between this stand and the two small forest bands farther below was too weak. Data used on single trees included their position, age, maximum injury heights, number and year of events, and DBH. In order to allow spatial visualization of the data in a GIS, tree coordinates were transformed into geo-objects and linked to their attributes from the database. Data elements were investigated with the Geostatistical Analyst extension (ESRI, 2008a) from the ESRI ArcGIS software (ESRI, 2008b) to examine spatial relationships between all sample points.

Following the procedure described in Johnston et al. (2003), skewed data were normalized as a first step. Trend analyses were then performed to identify directional influences (global trends), and data afterwards detrended using first- or second-order polynomials. Next, spherical semivariogram models and covariance clouds were used to analyze spatial autocorrelation and to adapt the number of lags and bin sizes. Cross validation of the measured points with the predicted points finally allowed the determination of mean prediction errors of the applied model. After consideration of the data, the Ordinary Kriging model (Johnston et al., 2003) was chosen for visualization of the continuous surfaces. We reviewed data from the five closest trees and at least two individuals for each of the four angular sectors to perform the interpolations for every single point in the interpolation sector.

4. Results

4.1. General aspects

In total, 191 trees were investigated in 19 horizontal transects, and the height and surface of 650 well-defined rockfall injuries were measured. As can be seen from Table 1, the large majority of sampled trees were

L. decidua (92.2%), whereas *P. abies* only accounted for 7.3%. *P. cembra* trees are scarce on the slope and only one individual was sampled for analysis: 167 trees (87.4%) were sampled along the 17 transects in the main forest stand, and 24 (12.6%) individuals were selected in the two horizontal forest bands located farther down (i.e. 12 trees per transect). Among the selected trees, 161 (84.3%) were sampled with increment borers, and 30 (15.7%) smaller individuals were felled with a handsaw. In total, 937 samples were chosen for analysis, namely 796 increment cores and 141 cross sections (Table 2). Tree age averaged 35.7

	Number of trees	%
Trees sampled		
L. decidua	176	92.2
P. abies	14	7.3
P. cembra	1	0.5
Forest stand		
Main forest stand	167	87.4
Small forest bands	24	12.6
Sampling method		
Increment borer	161	84.3
Handsaw	30	15.7

Table 1. Absolute and relative numbers of sampled species, their localization, and sampling method.

years (STDEV: 16.7 years), with the oldest tree showing 96 and the youngest only 12 increment rings. Tree age of the main forest averages 32.1 years, whereas the individuals situated in the two small forest bands are

	Number	%	Cores	Cross sections
Samples	937		796	141
Number of growth disturbances	2057	100	1656	401
Injuries	332	16.1	153	179
Traumatic resin ducts	1221	59.4	1028	193
Growth suppression	229	11.1	210	19
Growth release	56	2.7	55	1
Callus tissue	206	10	201	5
Compression wood	13	0.6	9	4
Reconstructed events	755		647	108

Table 2. Number of samples, growth disturbances (in absolute and relative numbers), and reconstructed events identified on increment cores and cross sections

older with 53.1 (upper band) and 68.5 (lower band) years, respectively.

4.2. Evolution of the stand and age distribution of trees

The analysis of the time series of aerial photographs covers the last 66 years of forest evolution on the study site. In 1941 (Fig. 3A), most surfaces of the investigated area were free of vegetation and the only trees existing were found in the lowermost part of the study site close to the bottom of the valley. Aerial photographs from 1981 (Fig. 3B) and 2004 (Fig. 3C), in contrast, illustrate the constant colonization of the study area with trees from the lower levels upward.

Tree-ring based data (Fig. 3D) confirm the age distribution and provide a more detailed picture of the afforestation history of the stand. It can clearly be seen that the oldest trees (>40 years) are clustered in the central lower part of the main forest stand. Towards the lateral borders of the stand, trees become gradually younger, with the youngest being identified at the upper end of the stand (average age of <20 years).

4.3. Bounce heights of rocks

In the field, the impact heights of the 650 injuries were recorded and grouped into classes of 20 cm. As can be seen from Fig. 4, injury heights show a normal distribution

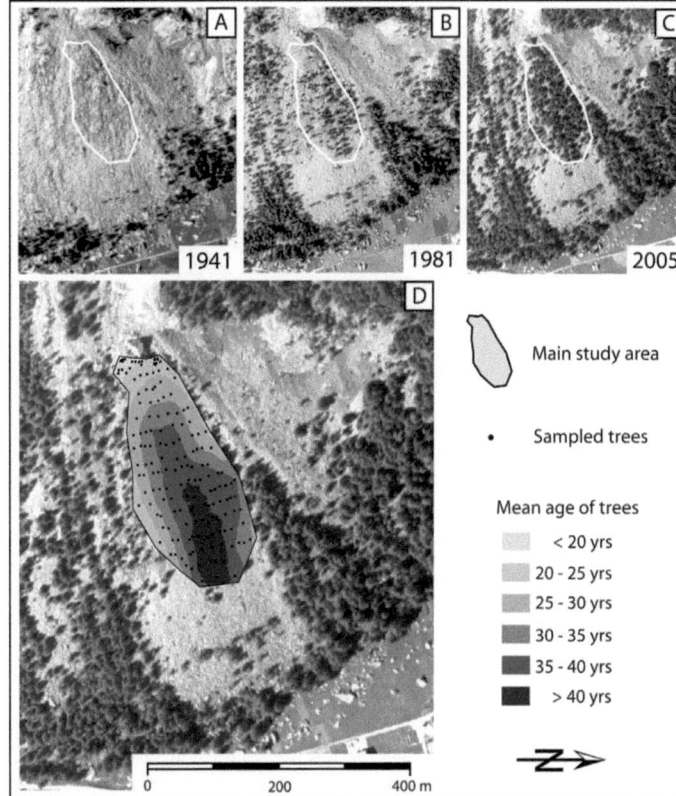

Fig. 3. Aerial photographs illustrating the colonization of the study area with trees in 1941 (A) 1981 (B) and 2004 (C). (D) Tree-ring data confirms the colonization process; oldest trees are situated in the lowermost central sector, youngest individuals grow on both sides of the uppermost of the forest stand (aerial photograph reproduced by courtesy of swisstopo (BA081119)).

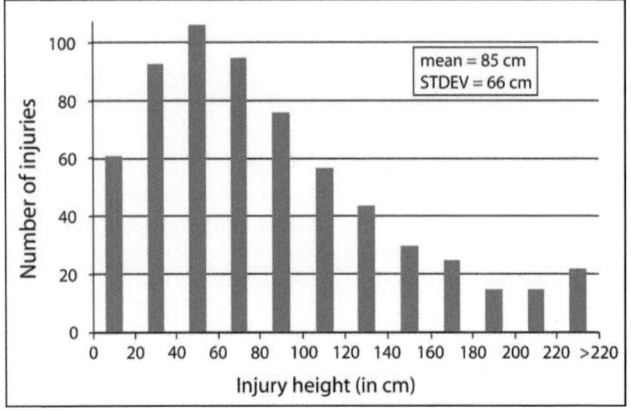

Fig. 4. Injury height distribution with a mean of 85 cm and a maximum of 450 cm.

with an average impact height located at 80 cm (SDTEV 66 cm). Most injuries (106 hits, 17%) were observed between 40 and 60 cm above ground level. More than two-thirds (67.4%) of all injuries were below 1 m, and only 22 scars (3.4%) were located above 2 m, with the highest injury identified at 4.5 m.

The spatial analysis and representation of maximum bounce heights of falling rocks was based on the uppermost injury per tree. The average maximum bounce height per tree for the entire stand is 126 cm for the entire slope and it accounts for 121 cm in the interpolated area (i.e., the main stand). As illustrated in Fig. 5, the spatial distribution of maximum bounce heights clearly indicates that the lowest values (<8 cm) are observed in the uppermost central part of the site where a huge boulder deviates or even halts rockfall. For the same reason, comparably low values are also observed farther down in the central sector where, in addition, the large number of neighboring trees has a certain shielding effect. In contrast, bounce heights become increasingly important towards the lateral boundaries of the forest stand, resulting in maximum average bounce heights >2 m in the southernmost segment of the study site.

4.4. Growth disturbances (GD) in trees affected by rockfall

In total, the analysis of the 937 samples allowed identification of 2057 GD and reconstruction of 775 rockfall events, resulting in an average of 3.95 events per tree and a mean recurrence interval of 9.0 years. As can be seen from Table 2, 332 injuries, or 16.1% of all GD, were directly visible on the samples (mainly on the cross sections). The most frequent GD observed were, however, tangential rows of traumatic resin ducts (TRD), which occurred 1221 times and represent 59.4% of all GD. Abrupt growth suppression was observed on 229 occasions (11.1%), whereas abrupt growth release could only be identified in 56 cases (2.7%). Callus tissue was present 206 times (10%), mostly in the segments of the rings bordering injuries. Finally, compression wood was only rarely present on the samples (13 cases; 0.6%).

4.5. Rockfall frequency, event years, and spatial distribution of activity

During the early colonization stage of the investigated forest stand, the number of samples was too small to allow comparisons of

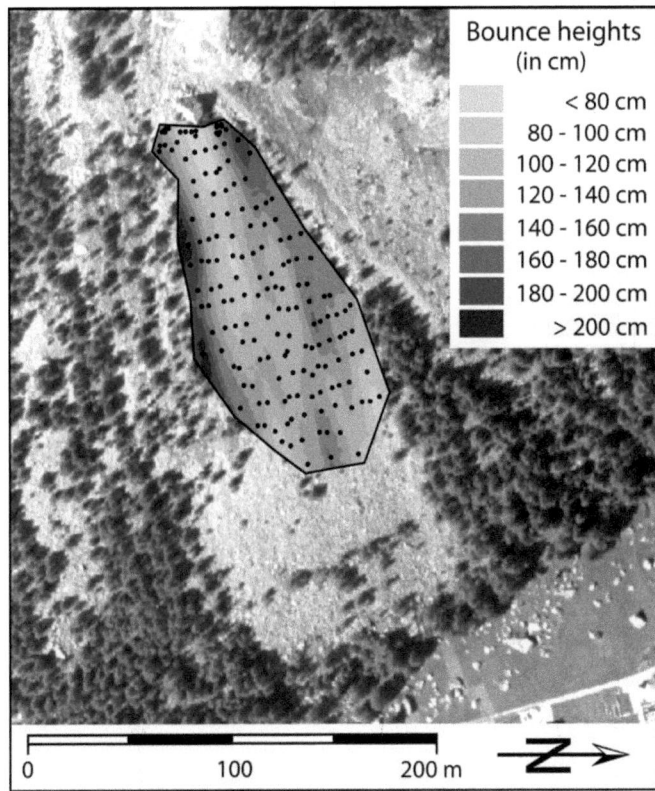

Fig. 5. Spatial distribution of maximum bounce heights. The big boulder identified in the uppermost part of the study site clearly alleviates bounce heights for the areas below. Highest values occur at the lateral boundaries (aerial photograph reproduced by courtesy of swisstopo (BA081119)).

past events and recent evolution of the rockfall activity. We therefore limited the reconstruction to the last 50 years (1957–2006), thus reducing the number of injuries to 745.

Rockfall activity on the slope is expressed as a rockfall "rate" and varies from zero to more than 6 events $m^{-1} y^{-1}$, with an average of 1.02 events $m^{-1} y^{-1}$ between 1957 and 2006 (STDEV: 1.2 events $m^{-1} y^{-1}$). As Fig. 6 illustrates, years with "no activity" are concentrated in the early colonization period when fewer trees were present for analysis. The last year with apparently no GD in the tree-ring series is noted in 1966. The absence of events in the early years occurred when the total diameter of all trees at breast height (DBH) was still low (2.3 m in 1957), whereas the exposed diameter of all investigated trees accounts for 38.2 m today. While the DBH increase was rather exponential between the 1950s and late 1980s, it became more gradual over the last 20 years.

As for event years, we observe several years with important rockfall activity in 1960, 1961, 1978, and 1995. In 1960, the rockfall rate attained 3.9 events m^{-1} and culminated in 6.6 events m^{-1} in the following year (1961), thus exceeding the average by a factor of 6.5. A less pronounced peak occurred in 1978 with a rockfall rate of 2.5. The latest peak in rockfall activity is noted for 1995, when 5.1 events m^{-1} were reconstructed from the tree-ring series. The 7-year moving average displayed in Fig. 6 does not show any significant trend in rockfall activity and is primarily influenced by the years

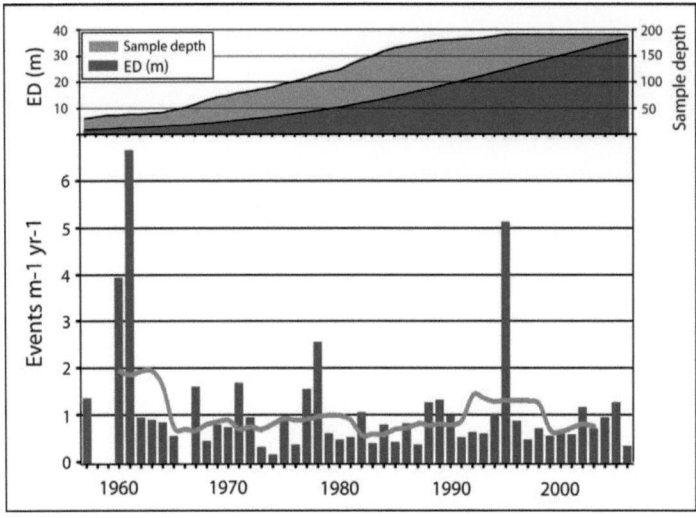

Fig. 6. Rockfall activity between 1957 and 2006 with the 7-year moving average (gray line), compared with the sample depth for the last 50 years (light gray) and the evolution of the ED (total DBH, dark gray).

with important rockfall activity in the early 1960s and in 1995.

Concerning spatial patterns, the highest rockfall activity can be observed at the lateral boundaries of the stand. The north-facing forest edge, pointing towards the main rockfall source area in a north-western direction, shows the highest activity with more than 4 events $m^{-1} y^{-1}$. In contrast, much less activity seems to be in the central area of the lowermost part of the study site, as trees are protected by their neighbors growing further above in the stand. The rockfall rate in the least active areas does not even attain 0.75 events $m-1 y^{-1}$ (see Fig. 7).

5. Discussion

In the present study,191 tree-ring series from *Larix decidua* Mill., *Picea abies* (L.) Karst., and *Pinus cembra* L. were investigated to reconstruct rockfall behavior over the last 50 years. In total, 796 cores and 141 cross sections were analyzed, yielding data on 937 samples. Based on the anatomy of the scars (i.e. tangential rows of traumatic resin ducts, presence of callus tissue, eccentric growth, formation of compression wood and abrupt changes in growth), 2057 GD were determined for these samples, resulting in 755 reconstructed events. Trees with a DBH N15 cm were exclusively sampled with increment borers. As the sampling was performed as close as possible to the wounds, GD were in general well pronounced and could be detected without any difficulty.

The reconstruction of tree ages and forest development was comprehensively confirmed by the visual analyses of the aerial photographs. Both methods reveal an afforestation that started in the central part at the lower levels of the study site and gradually moved farther up over the years. Based on the aerial photographs, apparently the reconstruction of rockfall activity over the last five decades appreciably corresponds with the time the forest stand has existed on the investigated slope. The photographs furthermore indicate that afforestation did not start at the study site exclusively but occurred on the entire slope (Fig. 8). The drastic change in the forest cover becomes very obvious when comparing the aerial picture

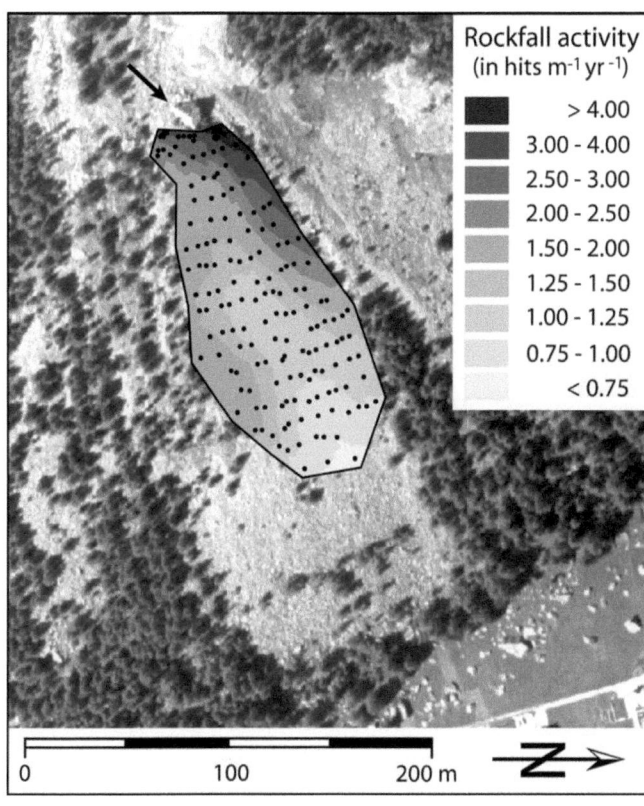

Fig. 7. Spatial distribution of rockfall activity. The highest activity is observed on the north-facing forest border, the lowest activity in the central part at the bottom of the study site. Notice the shielding effect of the big boulder (black arrow) at the upper end of the study site (aerial photograph reproduced by courtesy of swisstopo (BA081119)).

of 1941 with that of 2004. In 1941 (Fig. 8A), the slope was bare and the current-day forest almost nonexistent; whereas in 2004 (Fig. 8B), the entire slope (1 ha in surface) was covered with dense forest. Possible explanations for this quite rapid and ubiquitous afforestation could be (i) a drastic decrease in rockfall activity with the subsequent stabilization of the slope or (ii) the overall increase in mean annual temperatures as documented by meteorological data (Bader and Kunz, 1998) leading to better growth conditions for trees after the 1940s. Based on the low-resolution data and supporting information, it is, however, impossible to draw further conclusions as to the changes in rockfall activity, the existence of years with largely increased rockfall rates, or for the evolution of the stand.

Apart from the large-scale changes in the forest cover, the comparison of the aerial photographs also allowed identification of interesting small-scale changes. Fig. 8C and D show an extract of the departure zones where, sometime between 1958 and 1968, a large block (2000 m³) travelled for ~100 m from the frontal area of the departure zone to its current position at the uppermost limit of the study site. The aerial photographs also illustrate that the event completely destroyed the upcoming forest. Based on the tree-ring reconstructions and the increased rockfall activity in 1960 and 1961, we believe that the block was released during this period as rockfall increased.

The presence of this huge boulder does not only affect the trajectories of rockfall by

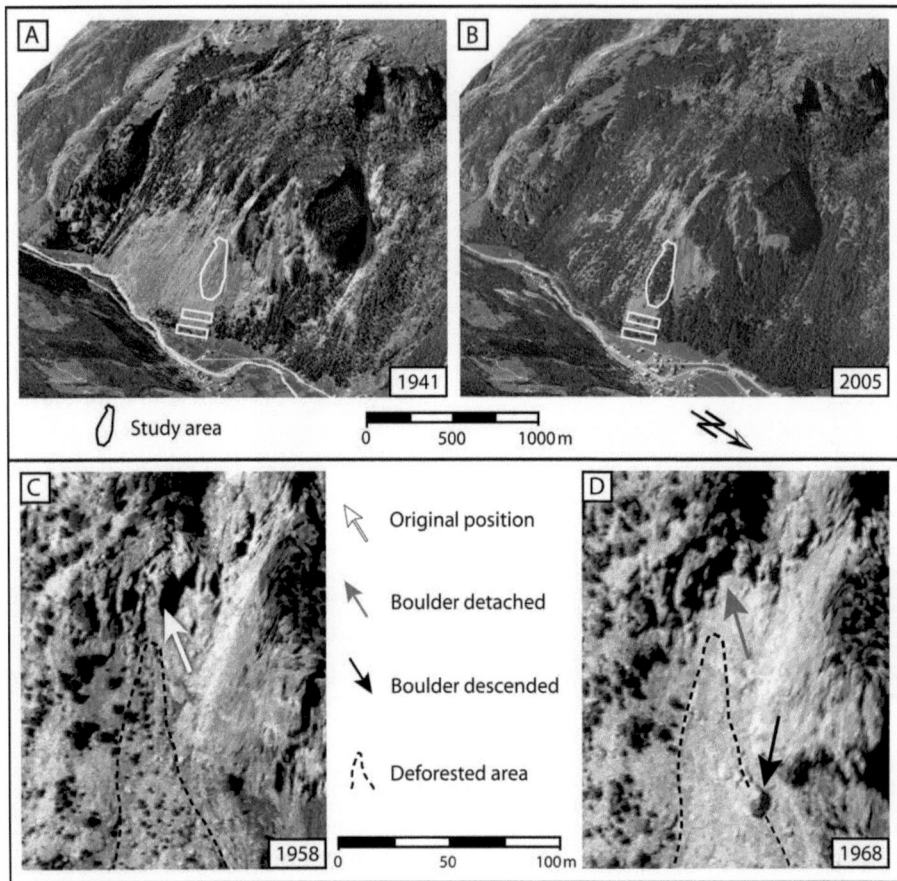

Fig. 8. Aerial photographs of the larger study site region draped over a DEM. (A) An unforested slope in 1941. (B) Identical view in 2004 when nearly the entire slope was covered with forest. (C) Detailed view of the detachment area in 1958 with the big boulder in situ (white arrow). (D) Identical view in 1968; the boulder has toppled down and a previously existing stand has been destroyed (aerial photographs reproduced by courtesy of swisstopo (BA081119)).

deviating or stopping rocks, but it also provides a shielding effect for the areas located in the fall line farther down the slope. This protective effect may also explain the abundant appearance of a dense forest at the study site, whereas trees are still missing north of the investigated stand and only occur in the form of a much less dense forest stand to the south of the study site.

The forest stand analyzed in this study was quite young with an average age of 35.7 years, as compared to other dendrogeomorphic studies covering several centuries (Stoffel et al., 2005b; Perret et al., 2006). The recent age of the individuals, however, simplified the sampling and the visual analysis of the samples in the laboratory, as we could deal with comparably small trees showing relatively few tree rings. A young tree age furthermore reduces the occurrence of hidden injuries, as there was not enough time for a wound to heal completely, thus making the scar undetectable when inspecting the stem's exterior (Stoffel and Perret, 2006). As we reconstructed a large number of events for a period of only five decades, we could provide a very extensive and dense data set that allowed a more detailed analysis of changes, with a much higher resolution in time and space.

We should also mention that the reconstructed number of events at the site has to be seen as a minimum frequency, because of the nature of the process itself and the method used to study rockfall. A falling, bouncing, or rolling rock can traverse a forest without hitting any trees and thus leave no marks at all. In addition, we did not take into account potential multiple events occurring in the same tree in the same year, as the approach used in this study does not allow differentiation between a single rock leaving multiple scars on a tree and several rocks hitting a tree at different times of the year.

At a more methodological level, we note that reactionwood was only very rarely observed in the selected trees (13 samples; 0.6%). In contrast to other mass movement processes (such as debris flows, snow avalanches, or landslides), rockfall does not seem to favor the formation of reaction wood resulting from the rock impact. It seems very likely that the energy transfer between the falling rock and the tree has to be seen as the main reason for this different reaction, as energy is transferred in the very short time span of only a fraction of a second, i.e., between 2 and 6×10^{-3} s in the case of concrete walls (Rutz, 1999) and, moreover, concentrated at a single point of the tree's stem. Another reason for the scarce appearance of reaction wood is the sampling strategy. Most of the samples were taken with an increment borer, very close to each wound. As reaction wood generally occurs in the opposite direction of the injury, detection is unlikely in a position adjacent to it. However, this is no explanation for the low appearance rate of reaction wood in the cross sections, where reaction wood can be found only in 3% of the samples. It is also true that the young age of trees certainly favours the absence of reaction wood, as they maintain a certain flexibility, which allows them to bend laterally during collisions.

The assessment of past global rockfall activity was based on a "rate" that includes the DBH in order to counterbalance the effect of the increasingly larger target surface offered to individual rockfall fragments with time. Results are given with a yearly resolution and indicate a strong variability of activity over the years. On the one hand, a certain background rockfall activity is present during "normal" years attaining 0.5^{-1} event $m^{-1} y^{-1}$. On the other hand, some pronounced event years exist with rockfall rates of more than 6 events $m^{-1} y^{-1}$. This variability can most probably be explained by variations of external factors, such as exceptional freeze-thaw cycles (Gardner, 1983; Matsuoka and Sakai, 1999; Matsuoka, 2006), the rising of mean annual temperatures (Davies et al., 2001), or extreme precipitation events (Schneuwly and Stoffel, 2008).

The analysis of the meteorological data for the 1995 event year revealed some extreme precipitation events before the onset of the vegetation period. The meteorological stations of Grächen (8 km northwest of the study site) and Zermatt (20 km southwest) both recorded exceptional rainfall on 5 November 1994 with 67 mm and 94.4 mm, respectively. Grächen started recording precipitation in 1959 and lists only five one-day precipitation events with more rainfall, whereas precipitation data of Zermatt shows only one event with more rainfall since the beginning of systematic recording in 1892. The local meteorological station in Saas Balen started recording in December 1994 and registered the second strongest precipitation (89.8 mm) event ever since at the 24. April 1995 (Schneuwly and Stoffel, 2008). It has to be mentioned that the meteorological conditions in this alpine region can significantly vary on a very local scale. Nevertheless, data shows two extreme rainfall events very close to the study site that could have caused (directly or indirectly) the exceptional rockfall activity recorded in the 1995 tree rings.

The outstanding rockfall activity in 1960/61 cannot be explained with exceptional precipitation events, as neither data from Grächen nor Zermatt indicate heavy rainfall events during the relevant period. In contrast, data from the Swiss Seismological Service note a magnitude 5.3 earthquake (Mercalli intensity VIII, 12 km depth) that would have occurred in Brig (18 km northeast of the study area) on 23 March 1960. There was no stronger earthquake in the Valais region ever since (ECOS, 2008) and we therefore believe that intense rockfall activity in 1960/61 would be the result of this seism.

The area with the highest rockfall activity is located in the north western sector of the study site. These results reflect the geomorphic situation in the field, as the most active area of the study site is located closest to the main source area of rockfall and does not have any trees farther above that could protect them. The generally lower rockfall rate at the southern boundary can be explained by the shielding effect of the big boulder located at the top of the stand and by the protection effect of trees located closer to the main rockfall source area. The least activity can be found at the very bottom of the study site, illustrating the substantial shielding effect of the trees growing farther above.

Finally, the yearly DBH increase was interpolated so as to obtain data on the growth conditions of trees on the study site. As a general rule, trees primarily increase their height in the first years before they increase their width (Kramer and Akça, 1995). At a certain stage, DBH increase will drop as a consequence of age. At the study site, tree growth is, in addition, disturbed by rockfall, which is why interpolated data (Fig. 9) neither individually reflect the age distribution nor the rockfall frequency, but rather indicate a mixture of both parameters. In the upper part of the study area, the interpolated yearly DBH increase nicely matches the pattern of the rockfall frequency (see Fig. 8). Growing conditions are the worst in the north western sector of the study site, and the DBH increase is apparently hampered by the very high rockfall activity. In the lower part of the main forest stand, both parameters seem to influence the conditions of growth: The largest DBH increase can be identified in a sector with rather a low rockfall activity and a moderate tree age.

6. Conclusion

Tree-ring analysis of 937 samples from nearly 200 trees allowed reconstruction of 755 rockfall events. The rather young mean age of only 36 years resulted in a dense data set that allowed detailed spatial analysis. Thus, spatial distribution of maximum bounce heights and rockfall activity could be assessed. While both of the methods used in the field and in the laboratory are relatively time consuming, they proved to be a powerful tool furnishing essential informa-

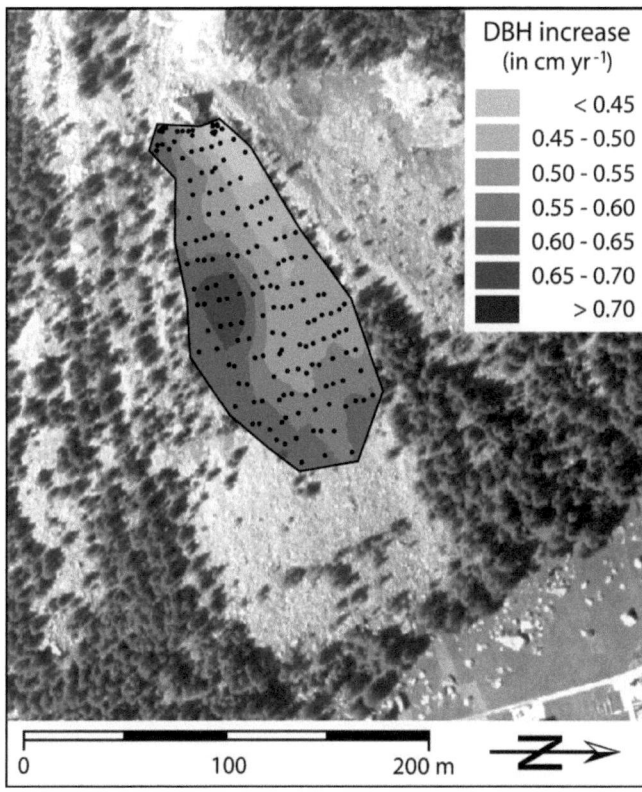

Fig. 9. Spatial analyses of the growing conditions using the yearly DBH increase as a reference. Worst growing conditions can be found in the areas with the most intense rockfall activity (aerial photograph reproduced by courtesy of swisstopo (BA081119)).

tion about the past and present behavior of rockfall on a forested slope.

Acknowledgement

The authors would like to thank Dr. Michelle Bollschweiler for her assistance in the field and the laboratory, as well as Oliver Hitz for his helpful comments on wood anatomy. We are also grateful to Lautaro Correa for his exhausting support in the field and furthermore kindly acknowledge Moe Mynt for his intelligible instructions on the GIS. Finally, we would like to express our appreciation to the local administration and the forest warden, who always allowed us to work in their forest stand.

References

Azzoni, A., Barbera, G.L., Zaninetti, A., 1995. Analysis and prediction of rockfalls using a mathematical model. International J. Rock Mechanics and Mining Sci. 32, 709–724.

Bader, S., Kunz, P., 1998. Klimarisiken — Herausforderung für die Schweiz. Wissenschaftlicher Schlussbericht NFP31, VdF Hochschulverlag AG, Zürich. (In German).

Bearth, P., 1973. Geologischer Atlas der Schweiz 1:25000, Simplon (Atlasblatt 61). Schweizerische Geologische Kommission.

Bearth, P., 1980. Geologischer Atlas der Schweiz 1:25000, St.-Niklaus (Atlasblatt 71). Schweizerische Geologische Kommission.

Berger, F., Quetel, C., Dorren, L.K.A., 2002. Forest: a natural protection mean against rockfall, but with which efficiency? The objectives and methodology of the ROCKFOR project. Interpraevent 2002, Band, 2, pp. 815–826.

Bollschweiler, M., Stoffel, M., Schneuwly, D.M., Bourqui, K., 2008. Traumatic resin ducts in *Larix decidua* stems impacted by debris flows. Tree Physiol. 28, 255–263.

Bozzolo, D., Pamini, R., Hutter, K.,1986. Rockfall analysis — a mathematical model and its test with field data. Proceedings of the 5th International Symposium on Landslides in Lausanne. Balkema, Rotterdam, The Netherlands, pp. 555–560.

Braam, R.R., Weiss, E.E.J., Burrough, P.A., 1987a. Spatial and temporal analysis of mass movement using dendrochronology. Catena 14, 573–584.

Braam, R.R., Weiss, E.E.J., Burrough, P.A., 1987b. Dendrogeomorphological analysis of mass movement: a technical note on the research method. Catena 14, 585–589.

Bräker, O.U., 2002. Measuring and data processing in tree-ring research — a methodological introduction. Dendrochronologia 20, 203–216.

Coe, J.A., Harp, E.L., 2007. Influence of tectonic folding on rockfall susceptibility. American Fork Canyon, Utah, USA. Natural Hazards and Earth System Sci. 7, 1–14.

Davies, M.C.R., Hamza, O., Harris, C., 2001. The effect of rise in mean annual temperature on the stability of rock slopes containing ice-filled discontinuities. Permafrost and Periglacial Processes 12 (1), 137–144.

Dorren, L.K.A., Berger, F., Putters, U.S., 2006. Real size experiments and 3D simulation of rockfall on forested and non-forested slopes. Natural Hazards and Earth System Sci. 6, 145–153.

Douglas, G.R., 1980. Magnitude frequency study of rockfall in Co. Antrim, N. Ireland. Earth Surface Processes 5 (2), 123–129.

ECOS, 2008. Earthquake Catalog of Switzerland. http://histserver.ethz.ch/intro_e.html. Erismann, T.H., 1986. Flowing, rolling, bouncing, sliding, synopsis of basic mechanisms. Acta Mechanica 64, 101–110.

ESRI (Environmental Systems Research Institute), 2008a. ArcGIS. http://www.esri.com/ index.html.

ESRI (Environmental Systems Research Institute), 2008b. Geostatistical Analyst. http:// www.esri.com/software/arcgis/extensions/geostatistical/index.html.

Evans, S.G., Hungr, O., 1993. The assessment of rockfall hazard at the base of talus slopes. Canadian Geotechnical J. 30, 620–636.

Fantucci, R., Sorriso-Valvo, M., 1999. Dendrogeomorphological analysis of a slope near Lago Calabria (Italy). Geomorphology 30, 165–174.

Friedman, J.M., Vincent, K.R., Shafroth, P.B., 2005. Dating floodplain sediments using tree-ring response to burial. Earth Surface Processes and Landforms 30, 1077–1091.

Harp, E.L., Wilson, R.C., 1995. Shaking intensity thresholds for rock falls and slides: Evidence from 1987 Whittier Narrows and Superstition Hills earthquake strong-motion records. Bulletin of the Seismological Soci. of America 85 (6), 1739–1757.

Hungr, O., Evans, S.G., 1988. Engineering evaluation of fragmental rockfall hazards. Proceedings of the 5th International Symposium on Landslides in Lausanne. The Netherlands, Balkema, Rotterdam, pp. 685–690.

Gardner, J.S., 1980. Frequency, magnitude, and spatial distribution of mountain rockfalls and rockslides in the Highwood Pass Area, Alberta, Canada. In: Coates, R., Vitek, J.D. (Eds.), Thresholds in Geomorphology. Allen and Unwin, New York, pp. 267–295.

Gardner, J.S., 1983. Rockfall frequency and distribution in the Highwood Pass area, Canadian Rocky Mountains. Zeitschrift für Geomorphologie N.F. 27, 311–324.

Gruber, S., Hoelzle, M., Haeberli, W., 2004. Permafrost thaw and destabilization of Alpine rock walls in the hot summer of 2003. Geophyiscal Res. Lett. 31, L13504.

Guzzetti, F., Crosta, G., Detti, R., Agliardi, F., 2002. STONE: a computer program for the three-dimensional simulation of rock-falls. Computers and Geosci. 28 (9), 1079–1093.

Johnston, K., Ver Hoef, J.M., Krivoruchko, K., Lucas, N., 2003. Using ArcGIS Geostatistical Analyst. Environmental Systems Research Institute (ESRI). Redlands, CA, USA.

Kirkby, M.J., Statham, I., 1975. Surface stone movement and scree formation. J. Geol. 83, 349–362.

Kramer, H., Akça, A., 1995. Leitfaden zur Waldmesslehre. J.D. Sauerländers. Verlag, Frankfurt am Main, Germany. (In German).

Lafortune, M., Filion, L., Hétu, B., 1997. Dynamique d'un front forestier sur un talus d'éboulis actif en climat tempéré froid (Gaspésie, Québec). Géogr. Phys. Quat. 51 (1), 1–15 (In French).

Luckman, B.H., 1976. Rockfalls and rockfall inventory data; some observations from the Surprise Valley, Jasper National Park, Canada. Earth Surface Processes and Landforms 1, 287–298.

Luckman, B.H., Fiske, C.J., 1995. Estimating longterm rockfall accretion rates by lichenometry. In: Slaymaker, O. (Ed.), Steepland Geomorphology. Wiley, Chichester, UK, pp. 233–255.

Marzorati, S., Luzi, L., De Amicis, M., 2002. Rock falls induced by earthquakes: a statistical approach. Soil Dynamics and Earthquake Engineering 22 (7), 565–577.

Matsuoka, N., 2006. Frost wedging and rockfalls on high mountain rock slopes: 11 years of observations in the Swiss Alps. Geophysical Res. Abstracts 8, 05344.

Matsuoka, N., Sakai, H., 1999. Rockfall activity from an alpine cliff during thawing periods. Geomorphology 28, 309–328.

McCarroll, D., Shakesby, R.A., Matthews, J.S., 1998. Spatial and temporal patterns of Late Holocene rockfall activity on a Norwegian talus slope: lichenometry and simulation-modelling approach. Arctic and Alpine Res. 30, 51–60.

Okura, Y., Kitahara, H., Sammori, T., Kawanami, A., 2000. The effects of rockfall volume on runout distance. Engineering Geol. 58 (2), 109–124.

Perret, S., Stoffel, M., Kienholz, H., 2006. Spatial and temporal rockfall activity in a forest stand in the Swiss Prealps — a dendrogeomorphological case study. Geomorphology 74, 219–231.

Ritchie, A.M., 1963. Evaluation of rockfall and its control. Washington, DC: Highway Research Board, National Research Council, Highway Research Record,17, pp. 13–28.

Ruppen, P.J., Imseng, G., Imseng, W., 1979. Saaser Chronik 1200–1979. Rotten-Verlag, Brig, Valais, Switzerland. (In German).

Rutz, J., 1999. Block-Anprall auf Stahlbetonwände aus Steinschlägen, Lawinen, Murgängen und Überschwemmungen. Gebäudeversicherungsanstalt des Kantons. St. Gallen, Sankt Gallen, Switzerland. (In German).

Sass, O., 2005. Temporal variability of rockfall in the Bavarian Alps, Germany. Arctic, Antarctic, and Alpine Res. 37 (4), 564–573.

Schneuwly, D.M., Stoffel, M., 2008. Tree-ring based reconstruction of the seasonal timing, major events and origin of rockfall on a case-study slope in the Swiss Alps. Natural Hazards and Earth System Sci. 8, 203–211.

Schweingruber, F.H., 1996. Tree Rings and Environment. Dendroecology, Paul Haupt, Bern, Switzerland.

Schweingruber, F.H., 2001. Holzanatomie. Paul Haupt, Bern, Switzerland. (In German). Shroder Jr., J.F., 1978. Dendrogeomorphologic analysis of mass movement on Table Cliffs Plateau, Utah. Quaternary Res. 9, 168–185.

Shroder Jr., J.F., 1980. Dendrogeomorphology: review and new techniques of tree-ring dating. Progress in Physical Geography 4, 161–188.

Solomina, O.N., 2002. Dendrogeomorphology:

research requirements. Dendrochronologia 20 (1), 231–243.

Statham, I., Francis, S.C., 1986. Influence of scree accumulation and weathering on the development of steep mountain slopes. In: Abrahams, A.D. (Ed.), Hillslope Processes. Winchester, Allen and Unwin Inc., Sydney, NSW, Australia, pp. 245–267.

Stoffel, M., 2008. Dating past geomorphic processes with tangential rows of traumatic resin ducts. Dendrochronologia. 26 (1), 53–60.

Stoffel, M., Perret, S., 2006. Reconstructing past rockfall activity with tree rings: some methodological considerations. Dendrochronologia 24 (1), 1–15.

Stoffel, M., Lièvre, I., Monbaron, M., Perret, S., 2005a. Seasonal timing of rockfall activity on a forested slope at Täschgufer (Valais, Swiss Alps) — a dendrochronological approach. Zeitschrift für Geomorphologie 49 (1), 89–106.

Stoffel, M., Schneuwly, D., Bollschweiler, M., Lièvre, I., Delaloye, R., Myint, M., Monbaron, M., 2005b. Analyzing rockfall activity (1600–2002) in a protection forest — a case study using dendrogeomorphology. Geomorphology 68 (3–4), 224–241.

Stoffel, M., Wehrli, A., Kühne, R., Dorren, L.K.A., Perret, S., Kienholz, H., 2006. Assessing the protective effect of mountain forests against rockfall using a 3D simulation model. Forest Ecol. Management 225, 113–122.

Stokes, M.A., Smiley, T.L., 1968. An Introduction to Tree-ring Dating. University of Chicago Press, Chicago., IL, USA.

Strunk, H., 1997. Dating of geomorphological processes using dendrogeomorphological methods. Catena 31, 137–151.

Wiles, G.C., Calkin, P.E., Jacoby, G.C., 1996. Tree-ring analysis and Quaternary geology: principles and recent applications. Geomorphology 16, 259–272.

Zinggeler, A., 1989. Steinschlagsimulation in Gebirgswäldern. Modellierung der relevanten Teilprozesse. Diploma thesis, Geografisches Institut, Univeristät Bern, Bern, Switzerland. (In German).

♦♦♦♦♦

Chapter D

Synthesis

1 OVERALL DISCUSSION AND CONCLUSIONS

1.1 MAIN RESULTS

The present work aimed at elaborating new methods for the investigation of rockfall activity using tree rings. Thereby, the high potential dendrogeomorphic methods in rockfall research should be demonstrated. The main part of this thesis was developed in four papers, presented in Chapters B and C.

In the first fundamental study of Chapter B, the anatomic tree reaction following rockfall impact was analyzed at the height of the injury. The analysis of the samples revealed that both species investigated (*Larix decidua* and *Picea abies*) reacted with the same types of anatomic growth features. It was shown that the formation of tangential rows resin ducts (TRD) was the most prominent response, followed by the differentiation of callus tissue. The tangential extent of the growth reactions were similar for both species. In *Larix decidua*, TRD were present on 34% of the remaining circumference, while callus tissue was found on 4.2%. TRD in *Picea abies* trees were formed on 36.4% of the remaining circumference, callus tissue was present on 3.6%. The intra-annual position of reactions allowed the reconstruction of the seasonal behavior of rockfall, giving evidence of potential local rockfall triggers.

The investigation of the intra-annual appearance of TRD also revealed a seasonal shift of TRD formation with increasing tangential distance to the injury in both species. However, the analysis of the long-term appearance of TRD revealed major differences between the species, TRD on *Picea abies* trees could be found for more consecutive years than in *Larix decidua*. It could finally be shown that the intensity of tree response depends on the width of the wound, the season of injury, and on the age of the tree at time of impact.

In the second fundamental study, anatomic tree reactions after wounding were studied in the tangential and vertical directions, focusing on the distribution of TRD and reaction wood in *Larix decidua*, *Picea abies*, and *Abies alba* trees. Again, both growth features appeared in all species, but with very differing intensities. *Abies alba* formed most dispersed TRD (in tangential and vertical direction), followed by *Larix decidua* and *Abies alba* trees. *Abies alba* individuals sparsely differentiate TRD following rockfall impact. Occurrence of TRD at each position around an injury was investigated. It was shown that the probability of TRD is highest just above the wound, followed by the lateral neighborhood and the sections below the injury. Above the wound, the probability of TRD formation trees exceeded 90%

in *Larix decidua* and *Picea abies,* whereas TRD in *Abies alba* could be found in 50% of the samples. Similar to the results obtained in the first fundamental study, an additional vertical intra-annual shift of TRD formation could be observed in all species. The analysis of reaction wood again revealed major differences between the species. In *Larix decidua*, reaction wood was rarely present on only 8% of the injuries. In *Picea abies*, reaction wood was found on 50% of the injuries and even on almost 90% in *Abies alba*. Reaction wood mostly appears on the samples at the height of the wound, followed by the samples above and below. Interestingly, no correlation between spread of TRD and size of injury existed in our samples.

Chapter C demonstrated several practical applications of the outcomes of Chapter B. In the first study, past rockfall activity on a case-study slope was assessed. Results indicate a regular background rockfall activity in "normal" years and extraordinarily high activity in a few "rockfall" years. It can be assumed that exceptional triggers cause the additional rockfall during these event years. Then, the seasonal behavior of rockfall activity was assessed. 76% of all events occur during the dormant season of the tree (early October to the end of May), whereas less than 3% take place between mid July and early October. However, in years with major rockfall activity, i.e. in 1995 and 2004, a different intra-annual distribution of events could be observed, again suggesting the presence of specific triggers. The analysis of meteorological data indeed revealed exceptional heavy precipitation events with high rainfall intensity in both event years. In a last analytical step, the main source area of rockfall could be reconstructed using the orientation of the wounds on the tree stem. The reconstructed main direction of rockfall was consistent with the geomorphic situation in the field. The difference between the natural fall line and the reconstructed direction amounted to more than 13 degrees (°).

More practical applications were finally presented in the last paper of the present thesis. In a first step, the afforestation process of the case-study slope was reconstructed. Then, the overall bounce heights of the slope were assessed, indicating a mean height of 85 cm, while two thirds of all bounce heights do not exceed 1 m. The maximum reconstructed bounce height was 4.5 m. In a next analytical step, the data on bounce heights were spatially investigated, indicating highest bounce heights at the lateral boundaries of the forested zone with decreasing values towards the bottom centre of the stand. Then, the rockfall frequency was successfully reconstructed for the last five decades, averaging at a mean value of 1.02 events m^{-1} DBH yr^{-1}. The analysis revealed several major rockfall event years, particularly in 1960/1961 and 1995, with a maximum frequency of 6.5 events m^{-1} DBH yr^{-1}. Again, the rockfall frequency was spatially analyzed, showing highest activity at the forest boundary that is oriented towards the main rockfall source area and lowest values at the bottom centre of the stand. For both major rockfall years, potential triggers could be identified. On 23 March 1960, an earthquake with a 5.3 magnitude (Mercalli intensity VIII, 12 km depth, ECOS 2008) occurred in 18 km distance from the site, explaining the outstanding rockfall activity in 1960/61. This earthquake would also explain the detachment of a huge boulder above the study site, discovered by comparison of aerial photographs from 1958 and 1968. Meteorological data show two exceptional heavy local rainfall events in the dormant season between early October 1994 and end of May 1995, leading to growth reactions at the very beginning of 1995. Finally, the yearly growth increase of sampled trees was spatially analyzed in order to determine the growth conditions at the study site. Worst growing conditions can be found in the zone with the highest activity.

In the four papers summarized above, the potential of dendrogeomorphic methods in rockfall research could be demonstrated. In the fundamental part of Chapter B, pioneering results on the anatomic behavior following rockfall impact could be gathered, facilitating future rockfall studies using tree rings. In the following, more application-oriented Chapter C, numerous applications for practical use were shown, by reconstructing different rockfall parameters at a case-study slope. There exists no other method that allows similar investigations on rockfall behavior.

1.2 Dendrogeomorphology in rockfall research

1.2.1 Potential

Dendrogeomorphology has proven its potential already in many research areas. Tree-ring research has successfully been used to investigate debris flows, landslides, erosion processes, or avalanches. In rockfall research, dendrogeomorphology was only recently applied by Stoffel et al. (2005a, b, 2006a) and Perret et al. (2006). The present thesis established the fundamentals for further studies in this field and showed several applications of tree-ring analysis in rockfall research. This thesis has proven the potential of dendrogeomorphic methods by successfully reconstructing many rockfall parameters. The practical use of tree-ring based data on rockfall is presented in the following.

Acquiring knowledge on past frequencies allows for a correlation of rockfall activity with different parameters, such as meteorological or climatic conditions. Analyzing past activity patterns may advance general understanding of the rockfall process and help the prediction of future development under different conditions, such as warmer temperatures or a changed precipitation behavior. The determination of main rockfall seasons allows identification of general rockfall triggers, supporting decisions in risk assessment. The knowledge on major event years gives evidence of exceptional triggers, such as extreme weather events or earthquakes. Again, knowledge on extreme events is indispensable for an accurate risk assessment. Reconstruction of local bounce heights with high spatial resolution can support the planning of distinct rockfall protection measures. Valuable data for protection measures can likewise be provided by spatially resolved rockfall frequency or by the determination of main rockfall trajectories. Numerical models were often used to determine runout distances or bounce heights. However, model verification and calibration in natural hazards is very difficult as reliable data is normally missing. Tree-ring gathered data on rockfall parameters can fill this gap and help the calibration of rockfall models. Adjusted models then can be used at sites where tree-ring analysis is not possible.

1.2.2 Limitations

So far, the possibilities of dendrogeomorphology in the analysis and reconstruction of past rockfall processes have been described. However, there are several limitations of tree-ring analysis.

The most important restriction is most likely the imperative presence of trees. Many scree slopes remain non-forested due to unfavorable growing conditions, such as limited availability of water and nutrients or an altitude above timberline. Another limiting factor is the tree species present at the site. Not all species respond with the formation of TRD or reaction wood. As seen in Chapter B2, *Abies alba* trees react weakly, others, such as *Pinus cembra*, do not form TRD after mechanical impact at all. Finally, the length and quality of reconstructed parameters is strongly influenced by the age of disturbed trees. No long term frequency can be established on a site with mostly young trees. Moreover, forested sites used for dendrogeomorphic analyses should not be influenced by human activity. Otherwise, it can not be guaranteed that growth responses are caused by geomorphic processes and not by construction or forestry work.

However, geomorphic processes likewise must fulfill several requirements. Only one geomorphic process should be present at the site. Analyses of growth responses do normally not allow the identification of the causing process, it would not be possible to lead back a reaction on a specific event. The only exception would be the concurring presence of seasonally separated occurring processes, such as debris flow or snow avalanches or the investigation of processes that leave different anatomical signatures in tree rings (Stoffel et al. 2006b, Stoffel and Hitz 2008). Finally, the geomorphic process, in the case of this thesis rockfall, should not be too devastating. If large scale events regularly occur on a slope, trees are regularly removed and no suitable forest can develop.

1.3 IMPLICATIONS OF RESULTS

1.3.1 ROCKFALL TRIGGERS

As seen in Chapter A2, there exist numerous potential rockfall triggers. The present thesis allowed for the identification of different local triggers. As already stated by Gardner (1980), it is possible to distinguish between ordinary "background" rockfall activity and extraordinary triggers leading to outstanding activity. The intra-annual distribution of ordinary rockfall at the study site identified freeze-thaw cycles as the main rockfall trigger. This result is in good agreement with many authors (Luckman 1976, Douglas 1980, Gardner 1980, Matsuoka 1990, 2001, 2008, Matsuoka et al. 1997, Sass 1998, Matsuoka and Sakai 1999, Rovera et al. 1999, Hétu and Gray 2000, Braathen et al. 2004, Stoffel et al. 2005b, Stoffel and Perret 2006, Kariya et al. 2007, Hall 2007).

At the study site, two additional rockfall triggers could be identified, responsible for extraordinary activity in major rockfall event years. In 1960, an earthquake of intensity 5.3 triggered both, a high frequency - low magnitude, and a low frequency - high magnitude event. This result confirms the findings of other authors, who likewise already identified earthquake as potential rockfall triggers (Keefer 1984, 2002, Case 1988, Bull and Brandon 1998, Rodríguez et al. 1999, Marzoratti et al. 2002, Braathen et al. 2004, Sepulveda et al. 2004, Becker and Davenport 2005). In the year of the earthquake as well as in the following year, many small scale events occurred, resulting in the highest rockfall frequency for the reconstructed period (five decades). Additionally, the quake most likely initialized the departure of an enormous boulder (2000 m^3) above the study site. The rock came to a stop right on top of the study site, where it remained in position ever since (Fig. D1.1, Fig. D1.2). The examination of aerial photographs confirmed the release of the boulder between 1958 and 1968.

The second local trigger of extreme rockfall

Fig. D1.1 Study site with the rockfall generating cliffs on top, where the 1960 earthquake presumably triggered the departure of an enormous boulder of 2000 m³ (Photo courtesy by Michelle Bollschweiler).

Fig. D1.2 Position of the boulder before (white arrow) and after (black arrow) the earthquake with a traveling distance of 100 m (Photo courtesy by Michelle Bollschweiler).

activity at the study site is heavy rainfall. This result again is in agreement with observation of numerous authors (Rapp 1960, Bjerrum and Jorstad 1966, Luckman 1976, Peckover and Kerr 1977, Gardner 1980, Kotarba and Strömquist 1984, Butler 1990, Nyberg 1991, Sandersen et al. 1996, 1997, Sass 1998, Krautblatter 2003, Rosser et al. 2005, Decaulne and Saemundsson 2006). The unusual activity between the end of 1994 and early 1995 could be correlated with two intense local rainfall events. Similar results on rockfall activity were presented by Stoffel et al. (2005b), who investigated past rockfall activity at a site in a neighboring valley (Täsch, Valais, Switzerland), focusing on the intra-annual seasonality of events. Their results likewise indicate highest activity in the dormant season and extraordinary rockfall activity in 1995 (Fig. D1.3). As

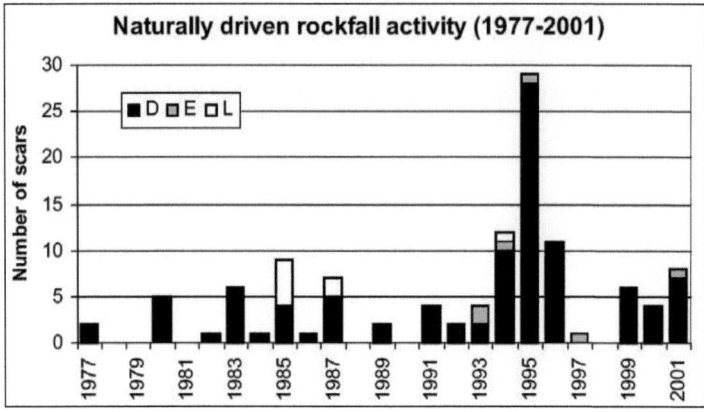

Fig. D1.3 Reconstructed rockfall events in a neighboring valley, likewise with exceptional activity in the dormant period (D) between the end of 1994 and early 1995 (Stoffel et al 2005b).

seen, rockfall is induced predominantly by very local triggers, the 1995 meteorological conditions however resulted in a regionally increased activity.

1.3.2 ROCKFALL AND CLIMATE CHANGE

Many authors predict a decrease of rock stability in a future greenhouse climate, leading to more rockfall activity. In general, the increased rockfall activity is seen as a consequence of the uplift of the permafrost boundary, resulting in cohesion lost and detachment of rocks. (Haeberli and Beniston 1998, Haeberli 1999, Schiermeier 2003, Gruber et al. 2004, Sass 2005, Krautblatter and Moser 2006). In practice, however, there is no clear evidence of increased activity in warmer climate so far. Due to a general lack of knowledge on past rockfall frequencies, there is only few data available to investigate potential correlations between rockfall activity and climate change. Stoffel (2005) compared 400 years of rockfall activity with corresponding temperature evolution, but did not detect any significant correlation. Perret et al (2006) however found a highly significant positive correlation between mean temperatures and the rockfall rate for the last century at their site. The differing findings of the authors can be explained by the ambiguous effect of rising temperatures. Perret et al. (2006) stated that increased winter temperatures potentially favor the occurrence of frequent freeze–thaw cycles and thus facilitates weathering and the production of rockfall fragments. Following Gruner (2008), the enhanced thermal contraction in cold winters results in increased joint widening and rock fracturing, leading to more loose material in the cliff and increased rockfall in the following season. However, he did not find a general correlation between temperatures and small rockfall events after investigation of several hundreds of past events. Data of the present study only reconstructed past activity for five decades, the high number of dated events during this "short" period of time however resulted in a very dense dataset. As seen in Chapter C, there exists no correlation of rockfall activity and temperature.

The melting of permafrost with resulting cohesion lost is not the only effect of a changing climate on small-scale rockfall activity. The modification of freeze-thaw patterns likewise leads to changed rock fractioning. As stated by Rovera et al. (1999), Matsuoka (1997, 2008) and Hall (2007), the comportment of freeze-thaw cycle is very complex and depends on many local factors, such as altitude, weather conditions, slope angle or slope orientation, leading to a very specific temperature pattern. A changed climate could therefore affect freeze-thaw behavior in many manners. A higher temperature regime would directly influence the number of freeze-thaw cycles. However, the effect could be ambiguous, increased temperatures can lead to more cycles at higher altitude, in zones where the water was mainly frozen so far. Likewise, higher temperatures could lead to fewer cycles at lower altitude, namely in areas that become simply too warm to regularly freeze water. Additionally, there exist several other factors that potentially affect the freeze-thaw pattern under changing climate, such as increased insulation or a modified precipitation regime. The isolating effect of the snow cover, for instance, controls surface temperatures and strongly influences the number and intensity of freeze-thaw cycles. A future seasonal shift of precipitation events from snow to rain would change the number of freeze-thaw cycles. An increasing number of intense rainfall events could lead to a more regular occurrence of major rockfall events. The latest assessment report on climate change by the United Nations predicts such a trend towards more extreme heavy precipitation events (IPCC 2007).

In order to determine future rockfall activity under changing climatic conditions, numerous parameters, ranging from global to local scale, must be taken into consideration. Changes of global climatic parameters finally modify very specific and interfering rockfall relevant parameters at the local scale. To determine the future rockfall frequency at a slope, all specific rockfall-inducing parameters as well as their functioning and interactions must be known. Global predictions on future rockfall are not possible, as equal changes do not necessarily result in similar reactions. This phenomenon likewise explains the differing findings of rockfall-temperature correlations at different study sites. However, climate change does affect many rockfall-relevant parameters and therefore has the potential to cause severe modifications in rockfall frequency or magnitude.

1.3.3 FUTURE ROCKFALL AT THE STUDY SITE

How does rockfall frequency change at the case-study site Saas Balen in a future climate? As seen, freeze-thaw cycles were identified as the main ordinary trigger while heavy rainfalls additionally have the potential to induce outstanding activity. This combination of freeze-thaw processes and heavy precipitation-related rockfall seems to be the main reason for alpine small-scale rockfall events, mainly occurring in late winter or early spring (Gruner 2008). Changes in freeze-thaw or precipitation patterns therewith have the potential to cause severe changes in frequency and magnitude. Both parameters are affected by climate change. Besides the knowledge of existing geological and climate properties at the site, future changes of different parameters must be known, namely: (i) future (seasonal) temperature evolution, (ii) future (seasonal) precipitation pattern, and (iii) future (seasonal) occurrence of extreme precipitation events. As described, rising temperatures for instance can both increase and decrease the number of freeze-thaw cycles. Therefore, all specific effects of locally changed parameters and their interactions must be investigated and assessed.

As there exists no data on the present freeze-thaw behavior and no data on local effects of climate change, no assured statement on future rockfall activity can be given. However, past activity pattern so far remained unchanged and revealed no correlation to climatic changes occurring during the last five decades. It is therefore likely that no short-term modification of frequency and magnitude must be expected at the site.

1.3.4 FOREST PROTECTION EFFECT

The presence of forest provides a protective effect against rockfall. Compared to non-forested slopes, falling fragments inside a stand seem to travel at lower velocity, rebound less high and come to an earlier stop (Meissl 1998, 2001, Perret et al. 2004, Dorren and Berger 2006 a, b, Stoffel et al. 2006a). However, quantitative data on the effective protection effect remain scarce, as there exist only very few direct comparisons (Dorren et al. 2006b). However, the outcomes of Chapter C2 confirm these results. At first sight, dendrogeomorphic methods do not allow for a comparison of forested with non-forested areas. However, the outer boundary of a forest records rockfall parameters existing outside the stand. Comparing parameters obtained from trees at the boundary with data recorded inside the forest could give interesting quantitative data on the protection effect of a forest. As seen in Chapter C2, data concerning rebound height or frequency clearly reveal differences between the forest boundary and areas within the stand. The general protection effect of the forest could clearly by demonstrated.

1.4 FINAL CONSIDERATION

Dendrogeomorphology has proven its unique potential in rockfall research. Numerous rockfall parameters could be reconstructed through the analyses of tree-ring series. In the first part of the present thesis, the fundamentals for several tree species with regard to practical applications were successfully elaborated. During the second part, a multitude of rockfall parameters could be reconstructed, such as past rockfall frequency, main rockfall source area, bounce heights, or main rockfall triggers. Spatial analysis of reconstructed parameters demonstrated the significant protection effect of forest stands against rockfall. Obtained data are of great value in practice. The evaluation of past rockfall frequencies allows for a more precise prediction of future activity. The intra-annual distribution of activity permits identification of main triggers, a dating of event years helps the determination of factors that can lead to extreme event years. Spatial analysis of frequency or rebound heights provides inevitable data for rockfall protection measures or can be used for model verification and calibration.

✦✦✦✦✦

2 REMAINING QUESTIONS AND FURTHER READING

This thesis revealed the wide range of possible applications of dendrogeomorphology in rockfall research and practice. However, there remain several open questions that require further research.

The theoretical fundamentals of tree response following rockfall impact could be established in a very complete manner. However, obtained results were restricted to several conifer species that represent the most abundant ones on alpine scree slopes. It is not known, what type, degree and distribution of reaction can be found in other conifer species or in deciduous trees. Moya et a. (2009) successfully dated rockfall injuries on *Quercus robur* L. (Pedunculate Oak). As no TRD are formed in deciduous trees, wedges of every injury had to be extracted to allow for a direct investigation of the wound. However, dating events with *Quercus ilex* L. (Holm Oak) was not possible as detectable growth rings are not correspondent to annual rings. As the theoretical knowledge on tree reactions of other species are mostly missing, dendrogeomorphic investigations would require research on the fundamentals first. Future dendrogeomorphic rockfall research would benefit from a larger fundamental database. Resulting data would facilitate and accelerate further applied dendrogeomorphic studies.

Another unknown is the maximum injury size trees can support. It is not known if there exists a mortal limit in wound width, length or area and in what degree this limit is species-dependent (Fig. D2.1). A further related interesting research topic is the capacity of trees to heal after wounding. So far, no data on wound closure or growth changes after natural rockfall impact exist (Fig. D2.2).

Fig. D2.1 *Example of a Larix decidua tree that survived a severe injury.*

Fig. D2.2 *Trees on rockfall slopes profit from efficient wound closing.*

The capacity of a tree to withstand severe wounding as well as the ability of efficient wound closing are of major interest in protection forest management.

During the present thesis, studies were mainly conducted at one case-study site. Similar studies at different other slopes could reveal interesting and useful outcomes. More data on alpine rockfall triggers would enlarge general knowledge on rockfall behavior and support evaluations in rockfall hazard and risk assessment. As seen, the 1995 rockfall event year could be detected on different slopes, triggering heavy precipitation thus operated not only at a local, but at least at a regional scale. Rockfall frequency reconstruction conducted at different sites could reveal local rockfall patterns. Exceptional activity could be correlated with extreme climatic or meteorological events, resulting in threshold values at which extreme rockfall could potentially occur. The better understanding of trigger functioning would likewise allow for a more detailed appraisal of trigger behavior under modified conditions. Therewith, improved predictions on future rockfall activity in a changing climate would be enabled.

Finally, dendrogeomorphology could be used to quantify the protection effect of forest against rockfall. Reconstructed frequency at the upper forest boundary corresponds to the frequency outside the forest. The protection effect could be assessed by analyzing occurring frequencies likewise inside the forest, for instance along different horizontal transects. Comparing the frequencies on top with the values obtained at following transects would, if the study was conducted on a uniform rockfall slope, reveal the protection effect of investigated forest. A first study in this area of research was performed by Moya et al. (2009), who found higher rockfall frequency at the higher transect (talus apex) than at the lower one.

✦✦✦✦✦

3 Bibliography of Chapter D

Becker, A., Davenport, C., 2005. Rockfalls triggered by the AD 1356 Basle Earthquake. Terra Nova 15, 258–264.

Bjerrum, L., Jorstad, F., 1966. Stability of rock slopes in Norway. Norwegian Geotechnical Institute 67, 59–78.

Braathen, A., Blikra, L.H., Berg, S.S., Karlsen, F., 2004. Rock-slope failures in Norway; type, geometry, deformation mechanisms and stability. Norwegian Journal of Geology 84, 67–88.

Bull, W. B., Brandon, M. T., 1998. Lichen dating of earthquake-generated regional rockfall events, Southern Alps, New Zealand. Geological Society of America Bulletin 110, 60–84.

Butler, D.R., 1990. The geography of rockfall hazards in Glacier National Park, Montana. Geographical Bulletin – Gamma Theta Upsilon 32(2), 81–88.

Case, W.F., 1988. Geological effects of the 14 and 18 August, 1988 Earthquake in Emery County, Utah. Utah Geological and Mineral Survey Notes 22, 8–14.

Decaulne, A., Sæmundsson, Þ., 2006: Meteorological conditions during slush-flow release and their geomorphological impact in northwestern Iceland: a case study from the Bíldudalur valley. Geografiska Annaler 88A(3), 187–197.

Dorren, L.K.A., Berger, F., Putters, U.S., 2006a. Real size experiments and 3D simulation of rockfall on forest slopes. Natural Hazards and Earth System Sciences 6, 145–153.

Dorren, L., Berger, F., Mermin, E., Tardif, P., 2006b. Results of Real Size Rockfall Experiments on Forested and Non-Forested Slopes. Disaster Mitigation of Debris Flows, Slope Failures and Landslides, pp. 223–228.

Douglas, G.R., 1980. Magnitude frequency study of rockfall in Co. Antrim, N. Ireland. Earth Surface Processes and Landforms 5(2), 123–129.

ECOS, 2008. Earthquake Catalog of Switzerland, http://histserver.ethz.ch/intro_e.html (as seen on 26 March 2009).

Gardner, J., 1980. Frequency, magnitude, and spatial distribution of mountain rockfalls and rockslides in the Highwood Pass Area, Alberta, Canada. In: Coates, D.R., Vitek, J.D., (eds.), Thresholds in Geomorphology. Allen and Unwin, New York, pp. 67–295.

Gruber, S., Hoelzle, M., Haeberli, W., 2004. Permafrost thaw and destabilization of Alpine rock walls in the hot summer of 2003. Geophysical Research Letters 31, L13504.

Gruner, U., 2008. Climatic and meteorological influences on rockfall and rockslides ("Bergsturz"). Interpraevent 2008, Conference Proceedings, Vol. 2.

Haeberli, W., 1999. Hangstabilitätsprobleme im Zusammenhang mit Gletscherschwund und Permafrostdegradation im Hochgebirge. Relief, Boden, Paläoklima 14, 11–30. (In German).

Haeberli, W., Beniston, M., 1998. Climate change and its impacts on glaciers and permafrost in the

Alps. Ambio 27(4), 258-265.

Hall, K., 2007. Evidence for freeze–thaw events and their implications for rock weathering in northern Canada: II. The temperature at which water freezes in rock. Earth Surface Processes and Landforms 32(2), 249–259.

Hétu, B., Gray, J.T., 2000. Effects of environmental change on scree slope development throughout the postglacial period in the Chic-Choc Mountains in the northern Gaspé Peninsula, Québec. Geomorphology 32, 335–355.

IPCC (Intergovernmental Panel on Climate Change), World Meteorological Organization (WMO), United Nations Environment Programme (UNEP)., 2007. Fourth Assessment Report on Climate Change, Synthesis Report.

Kariya, Y., Sato, G., Mokudai, K., Komori, J., Ishii, M., Nishii, R., Miyazawa, Y., Tsumura, N., 2007. Rockfall hazard in the Daisekkei Valley, the northern Japanese Alps, on 11 August 2005. Landslides 4, 91–94.

Keefer, D.K., 1984: Landslides caused by earthquakes. Geological Society of America Bulletin 95, 406–421.

Keefer, D.K., 2002. Investigating landslides caused by earthquakes – historical review. Surveys in Geophysics 23, 473– 510.

Kotarba, A., Strömquist, L., 1984. Transport, sorting and deposition processes of Alpine debris slope deposits in the Polish Tatra Mountains. Geografiska Annaler 66A(4), 285–294.

Krautblatter, M., 2003. The impact of rainfall intensity and other external factors on primary and secondary rockfall (Reintal, Bavarian Alps). Thesis, University of Erlangen-Nuremberg (Germany), Department of Geography.

Krautblatter, M., Moser, M., 2006. Will we face an increase in hazardous secondary rockfall events in response to global warming in the foreseeable future? In: Price MF, editor. Global Change in Mountain Regions. Duncow: Sapiens, pp. 253–254.

Luckman, B.H., 1976. Rockfalls and rockfall inventory data; some observations from the Surprise Valley, Jasper National Park, Canada. Earth Surface Processes and Landforms 1, 287–298.

Marzorati, S., Luzi, L., De Amicis, M., 2002. Rock falls induced by earthquakes: a statistical approach. Soil Dynamics and Earthquake Engineering 22, 65–577.

Matsuoka, N., 1990. The rate of bedrock weathering by frost action: field measurements and a predictive model. Earth Surface Processes and Landforms 15, 73–90.

Matsuoka, N., 2001. Direct observation of frost wedging in alpine bedrock. Earth Surface Processes and Landforms 26(6), 601–614.

Matsuoka, N., 2008. Frost weathering and rockwall erosion in the southeastern Swiss Alps: long term (1994-2006) observations. Geomorphology 99, 353–368.

Matsuoka N., Hirakawa, K., Watanabe, T., Moriwaki, K., 1997. Monitoring of Periglacial Slope Processes in the Swiss Alps: the First Two Years of Frost Shattering, Heave and Creep. Permafrost and Periglacial Processes 8, 155–177.

Matsuoka, N., Sakai, H., 1999. Rockfall activity from an alpine cliff during thawing periods. Geomorphology 28, 309–328.

Meissl, G., 1998. Modellierung der Reichweite von Felsstürzen. Fallbeispiele zur GIS-gestützten Gefahrenbeurteilung aus dem Beierischen und Tiroler Alpenraum. Innsbrucker Geographische Studien 28, Institut für Geographie, Universität Innsbruck, pp. 249. (In German).

Meissl, G., 2001. Modelling the runout distances of rockfalls using a geographic information system. Zeitschrift für Geomorphologie 125, 129–137.

Moya, J., Corominas, J., Pérez Arcas, J., 2010. Assessment of the rockfall frequency for hazard analysis at the Solà d'Andorra (Eastern Pyrenees). In Stoffel, M., Bollschweiler, M., Butler, D. R., Luckman, B. H. (eds.), Tree rings and natural hazards - a state of the art. Springer, Amsterdam.

Nyberg, R., 1991. Geomorphic processes at snowpatch sites in the Abisko Mountains, northern Sweden. Zeitschrift für Geomorphologie 35(3),

321–343.

Peckover, F.L., Kerr, J.W.G., 1977. Treatment and maintenance of rock slopes on transportation routes. Canadian Geotechnical Journal 14, 458–507.

Perret, S., Dolf, F., Kienholz, H., 2004. Rockfalls into forests: Analysis and simulation of rockfall trajectories — considerations with respect to mountainous forests in Switzerland. Landslides 1, 123–130.

Perret, S., Stoffel, M., Kienholz, H., 2006. Spatial and temporal rockfall activity in a forest stand in the Swiss Prealps—A dendrogeomorphological case study. Geomorphology 74, 219–231.

Rapp, A., 1960. Recent developments in the mountain slopes in Kärkevagge and surroundings, northern Scandinavia. Geografiska Annaler 42, 1–158.

Rodríguez, C.E., Bommer, J., Chandler, R.J., 1999. Earthquake-induced landslides: 1980–1997. Soil Dynamics and Earthquake Engineering 18, 325–346.

Rosser, N.J., Petley, D.N., Lim, M., Dunning, S.A., Allison, R.J., 2005. Terrestrial laser scanning for monitoring the process of hard rock costal cliff erosion. Quarterly Journal of Engineering Geology and Hydrogeology 38, 363–375.

Rovera, G., Robert, Y., Coubat, M., 1999. L'action des processus périglaciaires dans les badlands marneux des Alpes du Sud: l'exemple du bassin-versant du Saignon. Environnements périglaciaires (Association Française du Périglaciaire) 6, 41–52. (In French).

Sandersen, F., Bakkehoi, S., Hestnes, E., Lied, K., 1996. The influence of meteorological factors on the initiation of debris flows, rockfalls, rockslides and rockmass stability. In: Senneset, K., (ed.), 7th International Symposium on Landslides, Rotterdam.

Sandersen, F., Bakkehoi, S., Hestens, E., Lied, K., 1997. The influence of meteorological factors on the initiation of debris flows, rockfalls, rockslides and rockmass stability. Publikasjon - Norges Geotekniske Institutt 201, 97–114.

Sass, O., 1998. Die Steuerung von Steinschlagmenge durch Mikroklima, Gesteinsfeuchte und Gesteinseigenschaften im westlichen Karwendelgebirge. Münchner Geographische Abhandlungen Reihe B29, 347–359. (In German).

Sass, O., 2005. Spatial patterns of rockfall intensity in the northern Alps. Zeitschrift für Geomorphologie 138, 51–65.

Schiermeier, Q., 2003. Alpine thaw breaks ice over permafrost's role. Nature 424, 712.

Sepulveda, S.A., Murphy, W., Petley, D.N., 2004. The role of topographic amplification on the generation of earthquake-induced rock slope failures. In: Lacerda, W., Erlich, M., Fontoura, S.A.B., Sayao, A.S.F., (eds), Landslides: Evaluation and Stabilisation; Proceedings of the 9th International Symposium on Landslides. Balkema, Rio de Janeiro, pp. 311–315.

Stoffel, M., 2005. Spatio-temporal analysis of rockfall activity into forests – results from tree-ring and tree analysis. PhD thesis. Department of Geosciences, Geography, University of Fribourg. GeoFocus 12, 1–188.

Stoffel, M., Perret, S., 2006. Reconstructing past rockfall activity with tree rings: some methodological considerations. Dendrochronologia 24(1), 1-15.

Stoffel, M., Hitz, O.M., 2008. Snow avalanche and rockfall impacts leave different anatomical signatures in tree rings of *Larix decidua*. Tree Physiology 28(11), 1713–1720.

Stoffel, M., Schneuwly, D., Bollschweiler, M., Lièvre, I. , Delaloye, R., Myint, M., Monbaron, M., 2005a. Analyzing rockfall activity (1600-2002) in a protection forest – a case study using dendrogeomorphology. Geomorphology 68(3–4), 224–241.

Stoffel, M., Lièvre, I., Monbaron, M., Perret, S., 2005b. Seasonal timing of rockfall activity on a forested slope at Täschgufer (Valais, Swiss Alps) – a dendrochronological approach. Zeitschrift für Geomorphologie 49(1), 89–106.

Stoffel, M., Wehrli, A., Kühne, R., Dorren, L.K.A., Perret, S., Kienholz, H., 2006a. Assessing the protective effect of mountain forests against rockfall using a 3D simulation model. Forest Ecology and Management 225, 113-122.

Stoffel, M., Bollschweiler, M., Hassler, G.R. 2006b.

Differentiating past events on a cone influenced by debris-flow and snow avalanche activity – a dendrogeomorphological approach. Earth Surface Processes and Landforms 31(11), 1424–1437.

Die VDM Verlagsservicegesellschaft sucht für wissenschaftliche Verlage abgeschlossene und herausragende

Dissertationen, Habilitationen, Diplomarbeiten, Master Theses, Magisterarbeiten usw.

für die kostenlose Publikation als Fachbuch.

Sie verfügen über eine Arbeit, die hohen inhaltlichen und formalen Ansprüchen genügt, und haben Interesse an einer honorarvergüteten Publikation?

Dann senden Sie bitte erste Informationen über sich und Ihre Arbeit per Email an *info@vdm-vsg.de*.

Sie erhalten kurzfristig unser Feedback!

VDM Verlagsservicegesellschaft mbH
Dudweiler Landstr. 99
D - 66123 Saarbrücken
www.vdm-vsg.de

Telefon +49 681 3720 174
Fax +49 681 3720 1749

Die VDM Verlagsservicegesellschaft mbH vertritt

Printed by Books on Demand GmbH, Norderstedt / Germany